WONDERFUL POWER

WONDERFUL POWER

The Story of Ancient Copper Working in the Lake Superior Basin

Susan R. Martin

Wayne State University Press Detroit

GREAT LAKES BOOKS

A complete listing of the books in this series can be found at the back of this volume.

PHILIP P. MASON, EDITOR
Department of History, Wayne State University

DR. CHARLES K. HYDE, ASSOCIATE EDITOR
Department of History, Wayne State University

Copyright © 1999 by Wayne State University Press, Detroit, Michigan 48201.
All rights are reserved.
No part of this book may be reproduced without formal permission.
Manufactured in the United States of America.

03 02 01 00 99   5 4 3 2 1

Library of Congress Cataloging-in-Publication Data

Martin, Susan R., 1947–
    Wonderful power : the story of ancient copper working in the Lake Superior Basin / Susan R. Martin.
       p. cm. — (Great Lakes books)
    Includes bibliographical references (p.   ) and index.
    ISBN 0-8143-2806-7 (alk. paper)
    1. Paleo-Indians—Superior, Lake, Region. 2. Indians of North America—Industries—Superior, Lake, Region. 3. Indian metal-work—Superior, Lake, Region. 4. Indians of North America—Superior, Lake, Region—Antiquities. 5. Copper mines and mining, Prehistoric—Superior, Lake, Region. 6. Copperwork—Superior, Lake, Region. 7. Copper implements—Superior, Lake, Region. 8. Superior, Lake Region—Antiquities. I. Title. II. Series.
E78.S87M37 1999
338.2'743'097749—dc21   98-45330

To the memory of
DR. JAMES BENNETT GRIFFIN
1905 – 1997

Contents

List of Figures   9

List of Tables   11

Acknowledgments   13

1.
An Ancient People Extracted Copper from the Veins of Lake Superior of Whom History Gives No Account

CHARLES C. WHITTLESEY, 1863

Introduction   15

2.
All Phenomena, Natural and Artificial, Material and Immaterial, Mundane and Celestial

WILLIAM H. HOLMES, 1919

Unique Qualities of the Lake Superior Basin   23

3.
What Has Particularly Recommended Their Labors to Me Is Their Love of Truth

EPHRAIM G. SQUIER AND EDWIN H. DAVIS, 1848

Outstanding Personalities   45

4.
Of Heaps of Rubble and Earth along the Courses of the Veins

Early Evidence of Copper Mining   81

5.
Implements, Ornaments, and Objects of Faith of Great Variety
WILLIAM H. HOLMES, 1919

Ancient Technological Practices   113

6.
They Must Have Been Numerous, Industrious and Persevering, and Have Occupied the Country a Long Time
CHARLES C. WHITTLESEY, 1863

Daily Life   139

7.
Wonderful Power
Trading Copper   183

8.
The Remarkable Enterprise and Acumen of the Natives Are Made Apparent
WILLIAM H. HOLMES, 1901

Challenges for the Present and Future   215

Appendix: Some Native Copper Artifacts of the Upper Great Lakes Region	227
Notes	253
References	261
Index	277

Figures

1. The western Lake Superior region
2. Distribution of copper-bearing geological deposits around the Lake Superior basin
3. The extent of early mining evidence, Keweenaw Peninsula, Michigan, mapped by Charles C. Whittlesey
4. The Ontonagon Boulder
5. An Isle Royale landscape, ca. 1890
6. John W. Foster
7. Josiah D. Whitney
8. Daniel Wilson
9. Charles C. Whittlesey
10. Newton H. Winchell
11. William H. Holmes
12. George A. West
13. Roy W. Drier
14. James B. Griffin
15. Aboriginal mining pits at 20IR80, Isle Royale National Park, Michigan
16. Hammerstone scatter, Isle Royale National Park, Michigan
17. Exposed copper vein within aboriginal pit at 20IR80, Isle Royale National Park, Michigan
18. Experimental copper mining, Isle Royale National Park, Michigan
19. Intentional alteration of mining hammers: pecked and rilled
20. Intentional alteration of mining hammers: partial-grooved and multiple partial-grooved
21. Variability in mining hammers
22. Copper bead cross-section
23. Paleoindian and Archaic Tradition sites and localities
24. Woodland Tradition sites and sites of indeterminate age
25. Geographical distribution of some linked prehistoric mortuary cults
26. Decorative designs on prehistoric copper implements from the Lake Superior region
27. Portions of the John T. Reeder Collection of prehistoric copper artifacts
28. Awl variability; fishhooks

29. Some awl forms, Houghton County, Michigan
30. Some variability in prehistoric bead strands
31. Celt variability
32. Some celt forms, Keweenaw Peninsula, Michigan
33. Crescent variability
34. Some crescent forms, Upper Peninsula of Michigan
35. Knife variability
36. A cluster of copper ornaments and associated implements, Keweenaw Peninsula, Michigan
37. Projectile point variability
38. Some variability in projectile points, Upper Peninsula of Michigan
39. Spatula and spud variability
40. Spatula and spuds, Ontonagon County, Michigan

Tables

1. Changes in glacial and postglacial lake phases and levels in the Upper Great Lakes since 12 KBP
2. Intentional modification by frequency, locality, and rank order: Isle Royale and Caledonia hammerstones
3. Stages of copper deformation during cold hammering
4. Chronology of culture-historical periods in the Lake Superior basin and nearby
5. Early radiocarbon dates in the Lake Superior basin in association with worked copper
6. Some Middle Archaic dates for copper from Wisconsin and northern Michigan

Acknowledgments

There are many friends whose support made this work possible. I want to thank the people at Wayne State University Press, especially Charlie Hyde, who encouraged me from the start. Warm thanks also to Arthur Evans, Janet Witalec, and Jennifer Backer of the WSU Press, for their patience and sound advice. My spouse, Dr. Patrick E. Martin, had great confidence in me and gave me essential emotional support. My colleagues at Michigan Technological University (MTU) gave great encouragement too. Terry Reynolds gave me the space and time I needed to finish this manuscript. My colleagues Mary Durfee, Dave Landon, and Larry Lankton commiserated about the lonely aspects of writing up a big project. Special thanks are due to two anonymous reviewers, and to others who read drafts of this work and offered vital commentary: John Anderton, Ted Bornhorst, Terry Childs, Caven Clark, John Halsey, Peg Holman, Pat Martin, and Tom Pleger. Many thanks also to Vonnie Lutzke and Anita Maki of the Department of Social Sciences staff.

Faculty members at Michigan Technological University provided technical assistance: Dr. Ted Bornhorst (Geological Engineering and Sciences), Dr. Mark Plichta (Metallurgical and Materials Engineering), and Dr. Margaret Gale (Forestry). Bob and Sandy Slater (Publications/Photo Services), Joe Pyykkonen (Publications/Photo Services), and Brett Huntzinger (Forestry) gave their advice and artistic talents to the project. I thank the many Michigan Tech students who worked so competently in the lab: Dulci Avouris, Julia Bailey, Hugh Barnes, Karen Caldwell, Dennis Leopold, Barbara Sieders, and John Stevens. And special thanks to my promising student helpers, Amber Kenny and Holly Martin.

The staff of the MTU Van Pelt Library's Copper Country Archives and Interlibrary Loan office were essential and all deserve a big pay raise. Special thanks to archivist Erik Nordberg. The staff of the Milwaukee Public Museum library helped answer biographical questions about George A. West.

Thanks to Doug and Sylvia Barnard, Caven Clark, Bill Deephouse, David H. Thompson, and Liz Amberg Valencia for conversations and insights about Isle Royale National Park. The staff and friends of the Ontonagon County Historical Society Museum were extremely willing to help me whenever I asked, and I especially value the friendship of Bruce Ruutila.

Other colleagues answered frequent e-mail inquiries, located source materials, provided late-breaking data, asked provocative questions, and offered critical commentary, assistance, and encouragement: Tyler Bastian, Bob Birmingham, Richard Bisbing, Carl Blair, Jim Bradley, Jan Brashler, Adrian Burke, Frances Duke, Katy Egan, John Franzen, Betsy Garland, Holly Gersch, Jane Grenville, David Harvey, Earl Hautala, Gwynn Henderson, Mike Heyworth, Pat Julig, Karlis Karklins, David Kennedy, Michael LaRonge, Martha Latta, Mary Ann Levine, Bill Lovis, Tim Mighall, Claire McHale Milner, Charles Moffat, John O'Shea, Tim Rast, Paddy Reid, Bill Ross, Doug Scott, Steven Shackley, Beverley Smith, Janet Speth, Cindy Stiles, Larry Sutter, Paul Thibadeau, and Matt Thomas. Friends from the Upper Peninsula chapter of the Michigan Archaeological Society provided site locations and site visits, access to collections, and encouragement for this work. Bruce and Dan shared information, inspiration, and friendship over many years. Many other friends allowed me to photograph their collections. I appreciate also the help of Mark Hill and Troy and Jill Ferone of the Ottawa National Forest.

My long-suffering family understood as I slogged away on this project: they taught classes for me, shopped, did laundry, walked the dog, hired me a housekeeper as a birthday gift, gave me freedom to work, and endured my outbursts of frustration. Thank you with love to Pat, Dan, and Holly; I'm looking forward to a lot of carefree time in Agate Harbor with you. I will always cherish the love and encouragement offered by my sister Mary.

Over and over again I found inspiration from the many people, from a variety of backgrounds, who share a profound curiosity about ancient copper mining and regional prehistory. Robert Burcher, David Hoffman, Barry Jens, Marshall Payn, and others engaged me in high-spirited debates about the identities of prehistoric miners. Finally, this book is particularly for the many individuals who regularly hunt the Keweenaw for copper relics. I hope that you'll begin to gain the riches of understanding that scientific archaeology can unlock, which are so much more precious than things alone.

1

An Ancient People Extracted Copper from the Veins of Lake Superior of Whom History Gives No Account

—CHARLES C. WHITTLESEY, 1863

Introduction

> The ultimate purpose of the archeologist working within his special field is not merely to classify and describe the antiquities, but to aid in acquiring and making available such full and intimate knowledge of all the phenomena of aboriginal culture as to render possible their accurate application to the elucidation of the American race and thus to the history of the human race as a whole.
> —WILLIAM H. HOLMES, 1919

The World of Prehistoric Copper-Working

Sitting here at a computer keyboard at the close of the twentieth century, it is somewhat hard to imagine the worlds of earlier millennia, elaborate worlds made of the activities of prehistoric people. But we can begin to know these strangers better through archaeology. In the Lake Superior basin of North America, the earliest people followed many unique ways of life over thousands of years and left behind smatterings of the objects and meanings of their collective lives within the archaeological record. Archaeology defines the *archaeological record* as the deposits of accidentally and intentionally preserved remnants of past cultures and human behaviors. This archaeological record is widespread beneath our feet, but its data are fragile and easily destroyed. The past is partially knowable through the efforts of archaeology, but the things and materials that comprise the archaeological record rapidly disappear if they are accidentally or carelessly dug up and scattered. The places they were deposited in the past are undergoing changes in the present. Developments along lakes and rivers, altering of natural waterways, expansion of farms and towns, conversion of open land to malls and highways—all have taken a large toll on the antiquities of North America. Public lands are, in part, protected by legislation from unmanaged destruction, but private lands are not.

These threats to understanding the past are part of the reason for this book. If the dangers to the full history of the region are better known, some of the record may be preserved, particularly by the actions of informed private citizens. This book is, first, an attempt to reach nonarchaeologists to tell them the story of the long human history of the Lake Superior basin and to enlist their help to protect its archaeological record for posterity. Archaeologists, sadly, know all too well what these threats are and have done their level best to counteract them.

This book is also an effort to collect critical scattered information about the use of ancient copper by prehistoric people and to gather and preserve extant data about this unique resource for the general readership. It is hoped that systematic review of the extant literature about copper working will reveal new patterns of understanding from past events. Despite its uniqueness, the significance of Lake Superior copper in prehistory is not well appreciated beyond peripheral commentary in general accounts of eastern North American archaeology. As a whole the region is relatively unstudied. It is considered somewhat remote from the well-publicized "centers" of ancient America: the mounds of the central river valleys. Yet from early postglacial times the region was occupied for thousands of years by gatherers, hunters, and fishers who discovered that worked copper made a great addition to an already complex material culture. This discovery predated the development of the mounds by many millennia. Applying the techniques of stoneworking, which were already hundreds of thousands of years old, the people acquired a useful substitute for workable stone and a source of fine ornamentation and decoration. The material also acquired a reputation for power, as far as the people were concerned, for it was believed to be the property of the mythical underground and underwater masters, the manitous, who controlled and dispensed it. Eventually it was considered to be a special material and was sought after in prehistoric trade across the eastern half of North America.

This fascinating story includes many standing puzzles because not everything is known of the specific ways in which people mined copper, worked the metal, and exchanged it for other materials. We infer and assume much that we do not completely understand about these processes. At the same time, more refined means of analyzing prehistoric materials are developing, and these are revealing richer information about the physical manipulation of native copper. Trace element studies for geographical provenance, assays of minute amounts of residual carbon for refined dating sequences, metallographic examination of manufacturing signatures and stresses from wear, and experimental mining and metalworking provide ancillary evidence to that lifted from the archaeological record. This book collects this new information and confronts standing puzzles in its light.

Every so often, usually on a fine summer day, visitors appear at my office door with an intense need to know something about prehistoric copper. Some of them have traveled thousands of miles to the Lake Superior basin's Keweenaw Peninsula to find out about ancient mining. They have been all over the area and yet have not been able to grasp it, to find it, or to understand it. In some individuals this frustration in the face of a need to know takes the form of an obsession. I think it is plain old unsatisfied curiosity blooming, and it is a feeling that I understand completely. I cannot recommend a comprehensive written source about prehistoric copper to them, because there is not one currently in print. Most of these seekers are well read and have already managed to find some of the technical literature about prehistoric copper. However, it is not always that simple, even for a professional researcher, to track down scattered archaeological information. One must have access to a good research library and a willing library staff to help find the more obscure and the more interesting work.

This is exactly what I did to research and write this book. Not everyone has the time, the inclination, or the other resources needed to conduct the search for relevant information alone, nor should they have to. What are professionals for, anyway, if not to communicate what they do to a curious and interested public? I have tried to direct my research to answer the questions that are most commonly asked about Lake Superior basin prehistoric copper. How ancient is its use? Where was it mined and how did the people learn to work the metal? How did they fabricate and use their tools and implements? Who were the people who used it, and why did they stop using copper? Where were their settlements and what were they like? What evidence for trade was there? Of which tribes or groups were they members?

Some people may notice that these are not the same questions that professional archaeologists ask about ancient copper users. Professional archaeologists, most typically, want to know what other professional archaeologists think about prehistoric copper! But I have found in my research that there is, even among the obscure resources of the professional community, no single place to find a general account of the prehistoric use of copper in the Lake Superior district. This volume is a beginning, and I expect that as soon as it is in the hands of the professional community, they will likely register their commentary and their corrections with me. If only it stimulates public curiosity as well as professional debate! Our discipline's ability to generate profound, almost obsessive curiosity is its greatest asset, and it extends our professional utility to society at large if we care to confront and satisfy that curiosity. Nothing could be more welcome, as far as I am concerned. Curiosity, then correction, after all, is the point.

All of science, by the way, works to revise itself. Science is a system of knowledge gathering whose rules include constant self-criticism as well as the search for disproof of given or standing explanations. Conclusions are reevaluated in the face of new data. For that reason alone, this is not the end of the story but the beginning. It is my hope that readers may join in the effort to know the prehistoric past more completely. Individuals can make an important difference, by documenting private collections, by working with professionals to add to contemporary records about the past, and by donating collections, site locations, or excavation records to public institutions for preservation.

The data of archaeology can be complicated. In addition, archaeology uses its own jargon as a shorthand that its practitioners speak and write to one another. In reality it is possible to express what archaeology has to offer in more straightforward ways. Professional jargon can be stiff, off-putting, and at its worst downright confusing. The strengths of professional insights, which result from a logical stream of inquiry and analysis, can be saved without burdening the reader with unfamiliar jargon. This I have tried to do here, so that the fascinating information about the past is accessible to the nonspecialist reader. There is great public interest in and demand for the stories of past histories that archaeology can reveal. This book may meet some of that demand. By arranging inquiries and answers according to the patterns of research logic that prehistorians have refined, it can introduce the interested reader to the structure of arguments that archaeological practitioners routinely follow. For those who wish to delve more thoroughly into archaeological literature for archaeologists, the suggestions for additional readings provided at the close of each chapter provide an initial toehold.

In places this book is extremely speculative. I have tried to identify those places so that the reader can distinguish for him/herself the more substantial from the softer conclusions. For example, some mechanical designs and materials of tool hafts (that is, handles/attachments) are well known and are based on bodies of evidence recovered from firm archaeological contexts within the region. Other hafting methods are speculative on my part, because they are based on hafts from adjacent geographical areas. Still others are based on known physical properties of materials, the ways in which physical forces are transferred, and added doses of imagination. These topics are addressed in an appendix, which presents a catalog of familiar artifact forms as well as conjectural hafts. It tries to give copper technologies a full context of utilization and to place them within the complexities of their original cultural settings.

Theoretical Disclaimer

Another of the difficulties that confronts the nonspecialist reader of modern archaeological reports is the tendency for archaeologists to belabor what appear to be rather serious theoretical points in their writing. Some of this rigor is forced by standards of modern scholarship, which generally require that hidden assumptions of causality and the details of theoretical constructs be thoroughly revealed. In the midst of this theoretical posturing it is important to remember the forgotten humans who created the record to begin with. So I have condensed my theoretical baggage to the minimum; in fact it is down to one sentence that I hope may be carried without excessive burden: *This book assumes that people of all cultures and times adjust their behaviors, beliefs, and material requirements in response to and in awareness of their physical and social surroundings.* Whether this statement is true is really not so important. The point the reader should remember is that I have written everything here as though it were.

Copper was only one small part of the total world of the ancient people of the Lake Superior basin. Unlike other technological revolutions based on materials such as iron, petroleum, and uranium, the use of copper did not remake the face and substance of the societies and cultures that acquired it. Nor did it revolutionize social relations and economic traditions. Quite the contrary. The archaeological evidence suggests that copper was a material that, while unique in some properties, fit in well with the cultural behaviors that preceded its discovery. Copper coexisted with rather than replaced other aspects of cultural adaptations. As has been stated elsewhere, in North America there was a copper age, but never a Copper Age. The copper workers of the Lake Superior basin did not alloy their copper with other metals nor did they focus on metals that fit today's market definitions of the precious or the valuable. Worldwide, the metal-using cultures that garner most attention today are those in which metalworking was very early, smelting technology was complex, alloys of today's precious metals were center stage, symbolic articles of great aesthetic appeal were produced, and possession and control of the articles denoted differential social and economic power. These conditions did not occur in Lake Superior basin copper-using societies. Instead, the societies of the lake environment maintained a stable long-lived adaptation to a complex and unique environment. Copper was not essential to their ways of life, just useful.

Two Perceived Challenges

The story of Lake Superior basin prehistory, particularly that which involves the exploitation of native copper, is so fascinating that it has attracted the attention of a contingent of folks who believe that Superior basin copper was mined by ancient Europeans and shipped to the Old World to provide materials for the European Bronze Age. I do not hide my professional conclusions here. To my knowledge, there is no evidence whatsoever for pre-Viking contacts between Europeans and North American natives. If this evidence is ever found, it will revolutionize our understandings of global migration and emergent navigation technology. Because I favor following the rules of science, those being constant critical evaluation in the face of new data, nothing would make me happier than to find this evidence. The search for this evidence can be in fact the same sort of search process that archaeologists use to examine other hypotheses. The trouble is there is no such evidence to date to examine, and the archaeological record is, so far, silent on this topic.

There are other real challenges to archaeology. Collectors with acquisitive goals are relentlessly taking apart the record of the past. Unlike museums and historical societies, which routinely operate to share the fascinating story of the past with the public and educate interested people in the many stories of past life, collectors are motivated by personal and individualized goals. And, I am sorry to say, they are very good at what they do. The record of the past is being destroyed because of their activities. Buried copper artifacts are particularly vulnerable because of the use of sophisticated and sensitive metal detectors to search for them. *It is impossible to overstate the tragic nature of this loss or the rapidly waning time that remains for us to preserve what is left of the prehistoric archaeological record.* It is also difficult to document how extensive the damage to our past really is. To me one of the most poignant and probably accurate of estimates comes from a former artifact collector who cares enough about the past to recognize the impacts of random collecting. His comment? The woods are no fun anymore, he told me recently. There are too many people out collecting, and you can see their trails everywhere, he claims. Not too many years ago, he said, a person could wander all day, all weekend, in the northern woods and see no sign of another person. Yet nowadays, everywhere you look, are the characteristic divots or turned-over soil layers that mark the activities of metal detectors. The woods are full of collectors, and the past is going away with them in the trunks of their cars.

I realize, and have been reminded many times, that metal detecting is considered by some people to be an innocent, healthy, and fun activity and that it does more good than harm. My response is to the point: we all must take responsibility for the impacts of our personal actions, fun or otherwise. To take apart the past for ourselves is to deprive others of its richness, and, with enough time and collecting, to deprive everyone of the possibility of learning about prehistory. It is that simple. The past is larger than we are, it is larger than the present, and it is not ours to consume or destroy in the name of innocent fun or otherwise.

What Is Included Here

The Lake Superior basin is a unique environment whose peoples made use of an uncommon natural resource, that being pure copper. Prehistoric native people discovered this copper thousands of years ago, and in part made their lives, their technologies, and their spiritual beliefs around it. The story of this metal and its uses by native people should be told to a wide audience rather than muttered about in dusty academic journals and obscure scholarly manuscripts. This book is based on the latest scientifically derived information that finds support in the widest range of comparative data. Chapter 2 introduces the unique qualities of the Lake Superior basin environment and places copper in its local geological context. Other important resources of the region are also introduced. In chapter 3, the reader meets some of the outstanding and unforgettable personalities who began systematic archaeological investigations in the copper region and learns of their triumphs and setbacks. This is essentially the story of the growth of archaeology as a science in North America. Chapter 4 describes the material evidence that first revealed the phenomenon of ancient mining and presents experimental data regarding the actual methods of bedrock mining in preindustrial times. In chapter 5, one finds a review of the range of technological practices that ancient copper workers perfected and employed and an examination of how modern metallurgists painstakingly reconstruct probable ancient crafts. Chapter 6 tells the story of changes in daily life during eight millennia of Lake Superior basin prehistory. This part of the story attempts to place copper in its proper context as one of several important resources that prehistoric people used. Chapter 7 is about trading copper and what motivated people to acquire it. It is about the ritual power that copper was believed to convey and how that power was exchanged between humans and their supernatural contacts. Chapter 8 identifies some challenges for the present and the future: site preservation, public participation, speculative archaeology, and fruitful research directions. The book closes with a catalog of artifacts, identifying tool mechanics and variable functions, and a list of source materials for reference and future reading. There is no prehistoric copper-working industry quite like that of the Lake Superior basin anywhere in the world. So here begins its story.

2

All Phenomena, Natural and Artificial, Material and Immaterial, Mundane and Celestial

—William H. Holmes, 1919

Unique Qualities of the Lake Superior Basin

> I had risen from my couch upon the sand, and after walking nearly a half a mile along the beach, I passed a certain point, and found myself in full view of the following scene, of which I was the solitary spectator. Black, and death-like in its repose, was the apparently illimitable plain of water; above its outline, on the left, were the strangely beautiful northern lights, shooting their rays to the very zenith; on the right was a clear full moon, making a silvery pathway from my feet to the horizon; and before, around, and above me, floating in the deep cerulean, were the unnumbered and mysterious stars—I returned to our encampment perfectly well in body, but in a thoughtful and unhappy mood. In fact, it seemed to me that I had visited the spiritual world, and I wished to return hence once more.
>
> —Charles Lanman

Introduction

It is impossible to overstate the impact of Lake Superior on its adjacent natural and cultural landscapes; the lake makes the weather, the length of the seasons, and the character of the microenvironments occupied by animals, plants, and people. It determines the color of the sky and the temperature of the air. Those who see it, hear it, and live near it are really never removed from its effects. Its great volume, surface area, and cold depths can be very treacherous. Battles with the lake seem pointless; given enough time it always wins. Boats, houses, shorelines, docks and piers, navigational aids, erosion protection—all the trappings of the human scale of events are routinely diminished by the actions of the water.

But on a geological timescale, the lake is simply the most recent manifestation of an unending, dynamic set of events that continually make and deform the earth. One of the toughest challenges facing any archaeologist

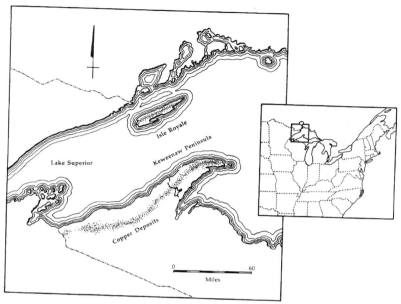

Fig. 1. The western Lake Superior region, indicating areas of copper deposits heavily exploited by native people. (Drawing by Patrick E. Martin; reprinted, by permission, from S. Martin 1994.)

is to achieve and then convey an understanding and appreciation of the vast time span of human life and culture. When this achievement is extended to the time span of earth processes such as bedrock formation, an even greater challenge is faced. But to comprehend the lay of the land and the geology of native copper, such a leap of scale must be made. One of the recurrent questions, which even the professional archaeologists and geologists ask each other, is how the early occupants of the Lake Superior basin found the veins of copper that they followed and exploited (fig. 1). As we develop an understanding of the genesis of native copper and its entrapment in regional bedrock, such questions become answerable. This chapter describes the landscape of the Lake Superior basin and the trap range:[1] the geological foundation, glacial alteration, and physiographic appearance of the region most exploited for bedded copper in prehistory (fig. 2).

Recurrent questions drive inquiry about native copper and its prehistoric uses. What characteristics of copper, and, overall, what geological processes are important in revealing its buried locations? What geological

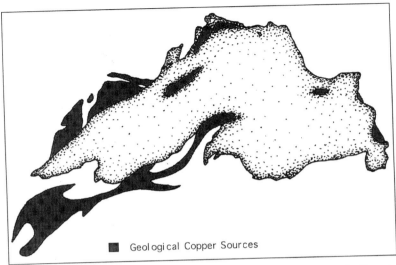

Fig. 2. Distribution of copper-bearing geological deposits around the Lake Superior basin. (Drawing by Brett A. Huntzinger, after Clark 1991:150; used with permission.)

processes account for copper's deposition across the glaciated portions of the upper Midwest? Did the native people depend primarily on mining embedded copper, or did they prefer collecting glacially scattered copper? What sizes of copper nuggets were most prized, and how were they related to geological processes? How chemically variable are the bedded deposits of copper? Are different lodes identifiable by trace elements unique to source locations? Does an artifact's elemental composition match that of a specific copper deposit?

Geology dictates some of the answers to questions about native copper and its prehistoric uses. For example, the surface distribution of copper due to displacement by glacial action is responsible, in part, for copper's particular transformation into a prehistoric commodity. It was apparently sought after and acquired through complicated and long-lived trading networks. Prehistoric natural resource economics, in common with commodity systems today, is a blend of natural as well as cultural distribution schemes. The processes prehistoric people used to extract and then manipulate the native copper were conditioned by the particular form and context of the copper's geological setting. Both depositional characteristics and the nature of the trap rock itself controlled, to some extent, to what uses the native copper could ultimately be put by aboriginal people. So, one resorts to geology.

Ancient Bedrock and the Genesis of Native Copper Deposits

This story is several billion years long and involves the processes that are responsible for the earth's crust and the formation of continental land masses. This story will be told in somewhat general geographic terms, dispensing, for the while, with specific localities. Likewise, one must accept the unfamiliar scales of error estimation that vast measures of time require. When something is as old as, say, 1047 million years, one must also accept an error factor of several million years' duration. The account below concentrates on two critical time periods: the formation of the copper deposits ca. 1050 million years ago (MYA), and the later glacial dispersal of copper during the Pleistocene epoch, ca. 1.8 MYA to about 10,000 years before present (BP). The discussion focuses on that portion of the Lake Superior basin where copper-bearing bedrock is exposed and was mined during prehistory: the trap range of the southwestern Lake Superior basin, the adjacent Keweenaw Peninsula, and Isle Royale.

The Precambrian rocks of the Lake Superior basin's western extent are some of the most ancient in the world. They make up in part what is called the Canadian Shield, and their formation and subsequent dynamic changes through cycles of deposition, erosion, and deformation laid the foundation of the Lake Superior landscape. By the end of the Archean eon of 4600–2500 MYA, major continents existed, made up of gneiss belts interspersed amongst neighboring greenstone belts. The greenstone belts contained extensive mineralization and were mostly formed "at subduction zones which were at or near the edges of continents" (LaBerge 1994:45). LaBerge suggested that in the would-be Lake Superior basin, the greenstone belts formed in an "ancient counterpart of an island arc environment" (45).

During the Early Proterozoic eon of 2500 to 1600 MYA, tectonic movement resulted in the collision of an active island arc with other continental crustal masses that were moving northward. The would-be basin area was "a gently rolling surface with a warm climate" (145). The first microfossils appeared in the Proterozoic, but the expansion of life on the surface was far in the future. Until ca. 1100 MYA, a lengthy span of erosion events occurred in the central part of North America.

The tranquility of this time ended with deposition of sands followed by numerous basaltic magma eruptions that lasted some 25 million years and the formation of the midcontinent rift system. The Lake Superior basin began to take shape when a deep-seated mantle plume resulted in doming, crustal thinning and spreading, and the formation of a rift, a down-dropped area bounded by faults due to a separation in the continental crust. This rift extended from Kansas toward the would-be Lake Superior region and through lower Michigan. The plume of rising heat created voluminous melting of the rocks in the earth's mantle. These magmas rose upward along weaknesses in the extending crust and erupted on the surface. Vast amounts of lava broke through and filled the deepening rift. Even some of the individual blankets of lava were enormous; the Greenstone Lava Flow of the Keweenaw, for instance, may be the largest individual lava flow on the surface of the earth (Bornhorst and Rose 1994:85). It was likely, from the appearance of the flows, that most were laid down on land surfaces and not underwater. Postvolcanism subsidence occurred when the thermal pulse that initiated doming, rifting, and lava eruptions diminished.

Intermittently, the outpouring of lava ceased and was followed by periods of sedimentary accumulation through erosion on the margins of the developing rift. Materials on the margin (earlier lavas) were broken down, transported by ancient rivers and streams, and redeposited as worn pebbles and sands in low-lying areas. These beds of conglomerate and sandstone were subsequently overlain by lava flows. In the exposed section of the lava flows on the Keweenaw Peninsula, about 20 conglomerate beds occur within approximately 200 individual lava flows. While once horizontal, these strata now are angled (the dip of the rocks) to the northwest under the lake basin. Northwest of the Keweenaw Peninsula on Isle Royale these same rocks are angled to the southeast under the lake basin. Thus, these rocks are part of a broad syncline.

As the lava that filled the rift cooled, and internal pressures lessened, volatile gases were trapped within the newly formed rock. These gases rose as bubbles to the surface of the flows and formed a frothy top. These vesicular tops of flows were porous and permeable and acted as receptacles for the subsequent precipitation of secondary minerals such as calcite, silver, and copper. In addition, some lava flow tops consisted of fragments of broken-up vesicular basalt, or breccia. The open spaces between the fragments also became sites of native copper deposition, precipitated from hot water (hydrothermal) solutions. Conglomerates and faults were also paths of mineral-bearing solutions, which moved through the bedrock and filled areas that were relatively porous. Both native copper and many other associated minerals were precipitated from these solutions. To prospectors, these other minerals were clues that copper deposition lay nearby. The copper was likely leached from volcanic rocks deep within the rift by the hydrothermal solutions and was then deposited in these same rocks at relatively shallow levels closer to the surface. The timing of the widespread deposition of copper postdated the formation of the bedrock strata by about 20–30 million years and occurred from about 1067 to about 1047 MYA (Bornhorst 1997).

The Keweenaw Peninsula contains the largest concentration of natural native copper known on earth. The primary exposure of the copper-bearing rock is a thin belt, 3–6 km wide, within erosion-resistant rock that forms the spine of the Keweenaw Peninsula. The belts extend into Wisconsin to the southwest, a distance of ca. 150 km, with additional exposure on the opposing syncline of the lake along Isle Royale, and sporadically elsewhere around the lake. The areas where there is the strongest evidence of prehistoric mining coincide with those of recent industrial-scale exploitation of copper. Most prehistoric exploitation came from four restricted areas: near Mass City in Ontonagon County, from the hillsides on both sides of the Portage Waterway, northeast along the trap from the Cliff Mine to Copper Harbor, and on the Minong Ridge at Isle Royale (fig. 3). Some evidence suggests that deposits at the Brule River and other locations in northwestern Wisconsin were also prehistoric sources for bedded copper (Thomas 1993).

Lake Superior copper occurs both as native (elemental) copper and copper sulfide ores. Some forms of the metal were more important than others to the prehistoric user. The geographically restricted sulfide ores were

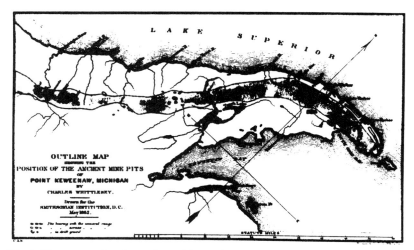

Fig. 3. The extent of early mining evidence, Keweenaw Peninsula, Michigan, mapped by Charles C. Whittlesey (Whittlesey 1863).

not important in terms of prehistoric exploitation. There was really no need for them; more than enough elemental copper was readily obtainable and usable. The "Indians never progressed to smelting technology" accusation that has been leveled at northern copper workers in the past overlooks the simple fact that there was really very little to be gained by smelting, given the presence of quantities of usable elemental metal.

The widely but irregularly distributed deposits were quite variable in size and shape, from fine dispersed bits of metal to pebbles, sheets, and boulder-sized masses (fig. 4). The deposits were very useful to prehistoric people if they were sheetlike or did not have to be divided. A fist-sized hunk could be hammered into a useful tool. A sheet could be folded, rolled, or reduced and turned into an object. However, detaching usable hunks from the largest masses was a technical difficulty with which even industrial-era miners struggled. The prehistoric people certainly did what they could to make use of these bonanzas; many of the large masses encountered during historic times, and even some today, show evidence of hammering by ancient people to worry off usable slabs of the material. The most useful deposits

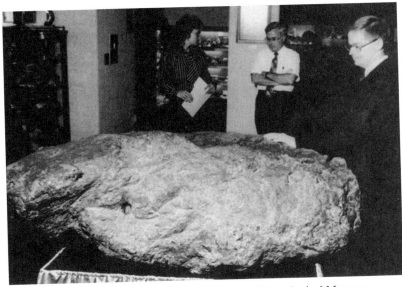

Fig. 4. The Ontonagon Boulder on display, Seaman Mineralogical Museum, Michigan Technological University, August 1987. (Photo courtesy of the MTU Archives and Copper Country Historical Collections, Michigan Technological University.)

were probably moderately thin sheets in fissure veins or copper masses of a size to shape into a tool without major reduction. Pebble-sized amounts occurred within glacial drift deposits over a wide area, as well as in surface deposits close to the bedded strata. Other forms of copper deposition, though plentiful, were too small to use for the production of artifacts.

To summarize, the copper deposits varied in form, in composition, in age, and in mode of original deposition. In addition, secondary deposition via erosion, water, and glaciation dislodged much material from its original locality and dispersed it across the north-central portion of the continent. Archaeological and geological researchers assessed copper's variable forms in order to understand more fully how it was acquired and used during prehistory.

Trace Element Studies and Sourcing of Archaeological Copper

Despite its relative purity (generally exceeding 99.9 percent), native copper contains inclusions of other elements in a variable range of proportions. These trace elements are exceedingly scant and are commonly measured on a parts per million (ppm) or parts per billion (ppb) scale. According to Wayman, the most frequently occurring trace elements are silver, arsenic, and iron, with a constellation of others including nickel, zinc, tin, cobalt, chromium, indium, and antimony (1985:75). Compositional studies of native copper vis-à-vis prehistoric metalworking began as early as the 1840s (Bastian 1961). The study and comparisons of the presence, absence, and amounts of accompanying elements can determine whether a piece of metal is an intentional alloy of copper and other metals. In addition, a particular profile of elements present in a specimen of copper, if matched with specimens of known geological or spatial derivation, could theoretically identify the place of origin of an unprovenanced specimen. These two themes, the attempt to identify alloy compositions and the attempt to match elemental profiles with geological sources, comprised most research about copper trace element studies. Bastian (1961) reviewed the uses of trace element studies in archaeology, focusing on North American copper.

Studying the differences between native copper and European smelted and/or alloyed copper began in the nineteenth century as an offshoot of an old argument about the sources of fancy copper artifacts from prehistoric American sites. Was this copper smelted, or perhaps was it actually not pure copper but an alloy such as bronze? Either finding would lend support to the notion that fancy archaeological metals came from European sources or influences. Other researchers wanted to distinguish early

European trade copper from native copper, which occasionally turned up together on sixteenth- and seventeenth-century sites. A host of archaeological researchers, beginning with Squier and Davis (1848), conducted this work, using a variety of methodological approaches to detect and measure trace element amounts. Today these methods, in addition to confirming that elaborate copper working was part of indigenous American technologies, now help to distinguish traded European copper from native copper found on historic-period archaeological sites (Fitzgerald and Ramsden 1988; Hancock et al. 1991).

Another research focus of archaeologists, geologists, and chemists examines the chemical signatures of different copper deposits. If these vary systematically, it is possible to match copper fragments and artifacts found archaeologically with their geological sources. The basic working hypothesis asserts that "when trace element impurities are present, the impurities are not of uniform occurrence in all native coppers, but may vary according to geographical factors and possibly through the geological mode of deposition" (Coghlan 1962:61). Such findings are potentially of great utility in understanding how and where copper was acquired and traded and what bedded sources of copper were important to prehistoric societies.

One obvious goal was to determine whether Lake Superior deposits provided all of the copper discovered archaeologically in the east, or whether a range of sources was used. There are a number of other native copper deposits in eastern North America that may have been used in prehistory—in the Appalachian Mountains from Georgia to New Jersey as well as in Nova Scotia. The degree to which these multiple sources were differentially relied upon for prehistoric supplies of copper is a subject of inquiry many decades old (Goad 1978; Hurst and Larson 1958; Levine 1995, 1996; Rapp, Hendrickson, and Allert 1990; Rickard 1934). Though Henrich[2] (1895) and Rickard (1934) posed the question earlier, Hurst and Larson's (1958) comparison of an archaeological specimen of copper from Georgia's Etowah site with geological specimens from Georgia, North Carolina, Michigan, and New Mexico was the first attempt to question the rote assumption that Lake Superior deposits provided all of prehistory's copper. Their goal was to differentiate the chemical signatures of their copper samples by geographical region. "A comparison of the trace element assemblages of these specimens provides a basis for deciding whether the Etowah copper is similar to that from the Great Lakes region or whether it more closely resembles native copper from a nearby source" (Hurst and Larson 1958:177). They discovered that while the three Lake Superior specimens they examined were very similar to one another in composition, the single Etowah sample most closely resembled the two geological specimens from Georgia sources. They concluded that it was no longer reasonable to assume that all prehistoric copper came from the Great Lakes. Goad (1978, 1980) and Goad and Noakes (1978) used spectrographic analyses, neutron activation analyses, and optical-emission spectroscopy to examine copper specimens, then constructed geographic clusters of copper source localities. This work concluded that favored sources of copper shifted during prehistory. In earliest times people relied heavily on the Lake Superior deposits, but by more recent times people developed control of local sources in the southeastern Appalachian Mountains.

Recently the availability of a growing data base of copper specimen trace element profiles makes it possible to compare more effectively the geological identities of copper sources utilized by early metalworkers (Levine 1996; Rapp, Hendrickson, and Allert 1990). Based at the University of Minnesota-Duluth Archaeometry facility, the data include trace element profiles from more than 1,200 specimens of native copper with known provenance and include samples of most known deposits north of Mexico. Systematic comparative analyses of archaeological specimens with copper specimens of known geological provenance may allow further assessment of patterns of copper acquisition, ancient exchange patterns, and shifts in procurement localities through time and between cultures. Some researchers suggest, however, that the extreme heterogeneity of native copper specimens and incomplete sampling of this heterogeneity limits the utility of trace element clustering for establishing provenance (Pletka 1991; Wayman 1991). Establishing differences between European trade metal and native copper has been somewhat more successful (Hancock et al. 1991; Wayman 1989).

The Natural and Cultural Landscape

Often we are so familiar with our daily surroundings that we take our local landscapes for granted. On any given day their implied permanence and stability may seem confirmable through everyone's individual observations. But according to a geological timescale, nothing could be further from the truth. In fact the landscapes around us are manufactured by the combined actions of nature and culture. The *natural landscape* itself is a misnomer. Wherever people have lived, their activities changed topography, changed water courses, and changed soil chemistry. They changed plant succession and changed habitat composition resulting in different distribution, numbers, and even behaviors of animals and plants. If humans are part of an ecological system, the resultant landscapes are cultural as well as natural products, both in the ways in which they are perceived and in their material and spatial arrangements. Obviously the scale of these alterations varies from culture to culture.

Prior to the occupation of North America by immigrants from Siberia and northern Asia about 20,000–15,000 years ago, the North American physical landscape was not influenced by people at all but was created by the double-sided forces of mountain building and erosion. In the most recent two million years, during the Pleistocene Epoch, these processes were helped along by periodic climatic shifts that resulted in global temperature changes and alterations in precipitation. These shifts generated expanding snowfields that perhaps up to twenty times extended over the central valleys of the current United States, some extending as far south as Illinois and Iowa (LaBerge 1994:244). The seasonal accumulation of snow, its transformation to ice, and its great weight produced a new phenomenon on the land, that of continental glaciation. It is accurate to say that these massive ice bodies flowed, melting at their warmer margins and adding snowfall at their colder margins, recycling water and suspended rock and sediment that created new formations far south of their bedrock origins. The glacial ice may have been as thick as 3000 m in the vicinity of the Keweenaw Peninsula (Bornhorst and Rose 1994:26). Prior to the scouring and depression of the region by the massive ice bodies, the Lake Superior basin was likely a lowland occupied by a north-flowing river (LaBerge 1994:240). The sediments of this valley were eroded and deposited far to the south by subsequent glacial action, excavating a bed that the lake eventually filled.

The long sequence of glacial advances and retreats lasted until ca. 9500 years ago in the northern Great Lakes. At the Gribben buried forest, Marquette County, Michigan, a stand of mature black spruce trees was buried as braided streams carrying heavy loads of glacial sediment changed course. In a few seasons' time flowing water quietly dropped loads of fine sands to a depth of ca. 8–10 m around the trees, killing them. This event occurred at approximately 9800 years BP (Drexler, Farrand, and Hughes 1983) at a time when the ice front was nearby, perhaps as close as 10–15 km to the north. Proglacial drainage fluctuations such as the one that affected the Gribben forest, as well as full-tilt advances and retreats of the ice front, continually rearranged topography and surface materials. The cumulative effects of these continental ice machines reshaped the central part of the continent many times, by altering drainage patterns, by creating new lakes and rivers, and ultimately by transforming old ones. Table 1 charts these changes over the last twelve thousand years for the northern Great Lakes.

Thanks to glacial erosion, geological deposits with copper inclusions were exposed at the ground surface in some places. LaBerge commented on the effectiveness with which glacial scouring produced "fresh bedrock" exposures in the Lake Superior region (1994:247). In addition, native copper was carried as a constituent of glacially deposited drift across the midwestern United States. Salisbury (1885) documented the geographic dispersion of

Table 1. Changes in glacial and postglacial lake phases and levels in the Upper Great Lakes since 12 KBP (after Anderton 1993; Eschman and Karrow 1985; Farrand and Drexler 1985; Hansel et al. 1985; Larsen 1985).

Time	L Superior	L Huron	L Michigan
Ice Events			
Readvance to Port Huron MI 12000 BP	Probable narrow ice-marginal lakes, episodic; record very indistinct	Ice and early L. Algonquin	L. Chicago and ice
Two Creeks interstadial after 12000 BP	Narrow ice-marginal lakes, episodic	Kirkfield low phase	L. Chicago Two Creeks low
Greatlakean advance 11800 BP	Ice filled	Ice and L. Algonquin	L. Chicago Straits deglaciated before 11000 BP
10400 BP	Early L. Minong precedes the readvance		
Marquette readvance 10000–9800 BP	Ice filled (east)		
Post-Marquette recession; Gribben buried forest at 9900–9800, associated with outwash from the Marquette readvance	West of the Keweenaw, L. Duluth and post-Duluth lakes (9800–9700). Rapid deglaciation of west with ice solid in eastern basin	L. Algonquin drops	L. Algonquin drops rapidly
		L. Stanley low shorelines submerged today	L. Chippewa low shorelines submerged today
Ice retreat at 9800–9500; retreat past North Bay, Ontario, about 9500 BP	Minong and sub-Minong lakes. Subsequent catastrophic surges from Lake Agassiz cause outlet incision and drop of Minong; Lake Houghton low (8000–7000 BP); these shorelines submerged today.	L. Stanley-Nipissing transition	L. Chippewa-Nipissing transition
Rising waters in Superior basin at 7500–4000 BP	L. Nipissing I rise	L. Nipissing I rise	L. Nipissing I rise
Water drop at 4000 BP	L. Nipissing II high	L. Nipissing II high	L. Nipissing II high
Water drop at 3200 BP	L. Algoma high; indistinct in many places; very close to current range	L. Algoma high	L. Algoma high ca. 3800 BP
3200 BP–2500 BP downcuts at Sault	L. Algoma drop to Sault level; also very close to current range	L. Huron drop	L. Michigan drop about 2500 BP
2200 BP–modern lake elevations with fluctuations	L. Superior ca. 183m	L. Huron ca. 177m	L. Michigan ca. 177m

Time differences per basin reflect both (1) actual time differences from southwest to northeast, and (2) the differential preservation of the glacial regression record across the lake basins.

drift copper to the state and county level as early as 1881, paying particular attention to confirmed discoveries and the contexts in which they were discovered. It has long been assumed that when these deposits were found to contain sporadic native copper, they served as prehistoric sources of the metal, at least in eastern Wisconsin, Illinois, and southern Michigan. Whether these were highly reliable and predictable sources, however, is debated (Rapp, Hendrickson, and Allert 1990). Salisbury suggested that the most abundant occurrences of copper within drift and lacustrine deposits were to be found in eastern Wisconsin and throughout the Kettle range (1885:46). According to his map of their widest distribution, glacial action was responsible for an east-west spread of greater than 700 miles and a southerly distribution as great as 600 miles from the source areas (47).

The glacial features in the copper-bearing trap range are the result of the waning of the most recent set of glaciations and consequent fluctuations in lake levels, from 11,800 to about 9500 years BP. The character of the local bedrock strata and its associated preglacial relief exercised a degree of control over the location of ice masses and drift deposits. Ice advances and regressions were fitful and sporadic by 11,000 BP. A pause point in the regression of glacial ice from the southern trap range resulted in the ponding of glacial meltwaters between the ice front and its terminal moraine to the south. This ponding, called Lake Ontonagon, filled the modern Ontonagon River drainage and deposited loads of distinctive red silt and clay there.

A series of later and short-lived recession-stage lakes covered the lands at the ice margins but rapidly waned as ice retreated northeastward. The evidence for their short lives is a series of very weakly developed stair-step beach configurations in the western part of the Lake Superior basin. At least one of these lakes, a falling stage of Lake Duluth, drained across the Keweenaw Peninsula through the Portage Gap while the Keweenaw Bay lobe of the ice front was still present to the northeast (Bornhorst and Rose 1994:32). The final recession of the ice was apparently rather quick by geological standards. For example, in the Ontonagon River drainage, radiocarbon evidence suggested that only about 800 years' time passed between the recession of the ice front and the final retreat of glacial meltwater lakes (Hack 1965).

Most of the surface of the Keweenaw Peninsula consists of glacially deposited till plain (Bornhorst and Rose 1994:32). The thickness of this material varies with underlying topographic characteristics and with the thickness of the ice sheet responsible for deposition. In Ontonagon County, the drift is mostly lake-bed sands and clays deposited by Lake Ontonagon. In the southern trap range, topography checked the ice action somewhat; there is morainic deposition bordered on the east by a band of outwash. Other water-borne glacial deposits are to be found sporadically in all areas of the trap range. These are of interest because they often contained substantial amounts of displaced copper nuggets and were exploited during prehistoric times for both copper and materials for stone tools.

Floral and faunal community changes accompanied the retreat of the proglacial lakes and the general climatic warming trend of postglacial times. Colonization of upland areas by species of spruce, pine, and some hardwoods began the development of a short-lived boreal forest, but gradual warming trends after 9500 BP influenced the entrance of more southerly species of both vegetation and animals into the region. This gradual replacement occurred as lake levels were attaining modern elevations. By about 3000 BP the distributions of modern floral and faunal communities were well established in the region, and by 2000 years ago modern lake levels had been reached throughout the Upper Great Lakes. Occasional climate-driven fluctuations in Lake Superior's levels occurred over the most recent 5500 years, best evidenced by buried soils at the southeastern shoreline of the lake (Anderton and Loope 1995:197–98).

A Composite View of the Pre-European Landscape

Though one does not expect landscape uniformity or stability over long reaches of time, the overall appropriateness of early-nineteenth-century analogs to the landscape of at least the prior 3000 years in the Lake Superior basin is justified. A virtual absence of farming, limited permanent settlement, limited impacts of logging and hunting/fishing, notable impacts related to use of fire and game fences, and habitual settlement at resource-rich or otherwise profitable locations; all are conditions that we rightly associate with non-European landscape uses, which dominated in the early nineteenth century throughout most of the Lake Superior shoreline. Thus the following description of that landscape derives both from nineteenth-century observations and present-day sources, and approximates the probable appearance of the region after the stabilization of postglacial lake stages about 2000 BP.

The trap range is rugged land dominated by abrupt relief, steep bluffs with rock outcrops, and dense conifer/mixed hardwood vegetation. Early geologists Foster and Whitney described its course:

> A trap range starts from the head of Keweenaw Point and runs west twenty miles; then, curving to the southwest, crosses Portage Lake near its head, and the Ontonagon river twelve miles from its mouth, and is thence prolonged into Wisconsin. Its length is more than one hundred and fifty miles; its width, from one to twelve. Between Iron and Presqu'-Isle rivers a spur shoots off in the form of a crescent, constituting the

Porcupine mountains. Another spur branches off from the main chain to the south, and is prolonged nearly parallel with it for twenty miles. This belt is made up of parallel ranges, presenting step-like or scalar declivities on the side opposite the lake, while the other consists of gradual slopes. (1850:34–35)

Sommers (1977:24–25) divided the general area into physiographic regions based on relief, elevation, and topography. Bordering the lake lies a very narrow Lake-Border Plain of low altitude, ca. 183–250 m above sea level (asl), and low relief, which follows the general contour of the present shoreline. Immediately inland from this plain (and in some cases directly adjacent to the shoreline) rises the Keweenaw Highland, a higher-altitude zone of abrupt relief underlain by Upper Precambrian volcanics: the trap range. Here altitudes reach greater than 500 m asl. Bedrock characteristics create the shape of the shoreline; "where igneous rocks prevail, the coast is finely indented; where the sandstones prevail, the coast is gently curved" (Foster and Whitney 1850:21). At Isle Royale, Lanman offered a succinct description of this zone: "Its hills have an altitude of four hundred feet, it is covered with forest, and has a bold shore—the northern side is bold and rocky, but the southern shore has a number of fine bays and natural harbors. The soil is barren—" (1978:133–34).

The near-coast mainland is drained by numerous short rivers with small drainages; their drop to the lake from the highlands creates numerous small waterfalls. They are marked by unstable flow and temperature. The north-flowing Ontonagon River is the largest and most mature drainage feature associated with the interior of the trap range; its four branches cover ca. 3600 sq km. Its drainage connects the range with the south-flowing Menominee River basin, which was extensively used as a transportation route during prehistory. Henry Rowe Schoolcraft described the Ontonagon in 1820 as he toured upstream to see the Ontonagon Boulder, the renowned three-ton mass of native copper:

> The lands along this river are generally rough and mountainous, until within three or four leagues of its mouth. Its waters have a reddish color[3] like those of the Arkansas, and are moderately turbid; among its forest trees pine and hemlock predominate, but its most remarkable character is the copper, which is found along its banks. . . . A broad river, with a gentle current,—winding course, and heavy wooded banks, with the dark green foliage overshadowing the water, rendering the first part of the tour delightful. (1970:172)

Later on, upriver by twenty or so miles, after two scorching days' travel, his enthusiasm for the local landscape was considerably reduced:

One cannot help fancying that he has gone to the ends of the earth, and beyond the boundaries appointed for the residence of man. Every object tells us that it is a region alike unfavorable to the productions of the animal and vegetable kingdom; and we shudder in casting our eyes over the frightful wreck of trees, and the confused groups of falling-in banks and shattered stones. Yet we have only to ascend these bluffs to behold hills more rugged and elevated; and dark hemlock forests, and yawning gulfs more dreary, and more forbidding to the eye. (1970:178)

Foster and Whitney were hardly more flattering regarding the interior of copper-bearing Isle Royale (fig. 5), claiming that its physical obstructions were the most formidable in the district: "So dense is the interwoven mass of foliage that the noonday sun hardly penetrates it. The air is stifled; and at every step the explorer starts up swarms of mosquitoes, which, the very instant he pauses, assail him" (1850:81). Visitors on a University of Michigan natural science expedition twenty years later were equally stunned by the harshness of that island:

> Had Dante lived on Isle Royal [sic] when he wrote the *Inferno*, he would without doubt have described the souls of one class of offenders

Fig. 5. An Isle Royale landscape, ca. 1890. (Photo, probably by W. W. Stockly or Alfred C. Lane, courtesy of the MTU Archives and Copper Country Historical Collections, Michigan Technological University.)

as condemned to be perpetually lost in a cedar swamp. To be compelled to flounder through the semi-darkness which always prevails there, to be buffeted by the thick bushes and struck in the face by the long, pendant lichens, to stumble over rotten trunks of prostrate trees and have the deceitful mosses let you down into the stagnant waters beneath and to feel the slimy water-weeds—suggesting nameless horrors. ([Anonymous] 1869:279)

The trap range falls within the Canadian Biome as defined by Dice, "that part of northeastern North America in which hardwoods form the climax and conifers of several kinds form several types of subclimaxes" (1943:503). The climax vegetation was a maple-birch-hemlock complex on moist acid soils, with local concentrations of pine, particularly in the southern Ontonagon River drainage, where sandy drier soils are common. "The entire country is rocky and covered with a stunted growth of vegetation, where the silver fir, the pine, hemlock, the cedar and the birch are most abundant" (Lanman 1978:128). The Canadian Biome is distinctively transitional between the conifers that dominate the boreal (northern) forests and the deciduous broadleaf forests of the south. It is an area of high species diversity of both southern and northern forms. Isle Royale is particularly endowed with boreal species because of the inescapable effects of Lake Superior's cold water there: "The shores are lined with dense but dwarfed forests of cedar and spruce, with their branches interlocking and wreathed with long and drooping festoons of moss" (Foster and Whitney 1850:81).

Prehistoric settlers in this region no doubt capitalized on the wide range of species available and based their adaptation to the area on profound knowledge of local microhabitats. Seasonally available foods such as fish, wild rice, game species, and maple sap provided the base upon which prehistoric cultures depended. The best sources of information about pre-European plant and animal uses come from the ethnohistorical and archaeological records of the region. Yarnell's (1964) study of the uses of plants by aboriginal cultures of the Upper Great Lakes identified 130 food species, 18 species used for beverages, 68 plants used as medicinal teas, 207 other plants used as medicines, 31 charm and ceremonial plants, 27 smoking plants, 25 dye plants, and at least 52 plants used for technological purposes. Though not all of these plants were available in the trap range, seasonal travels to other areas no doubt broadened plant use choices. The most important food plants in the north were probably maple (for sap), wild rice, berries, and tubers.

Cleland, studying the distribution of animal bones in archaeological context at the Juntunen site in Mackinac County, Michigan, identified 17 species of mammal, 16 bird species, 3 turtle species, and 18 species of fish which were probably used for food (Cleland 1966). This represented a minimum list of species available, because it derived from a single locality and emphasized warm-season food collection. Deer, bear, beaver, trout, whitefish, and sturgeon were among the most important species of animal food for native people in the region, at least in the thousand years preceding the appearance of Europeans.

The Presence of Lake Superior

The major factor in the local environment of the trap range is the presence of the largest freshwater surface in the world. The color of the sky and the intensity of the sunlight close to Lake Superior is very distinctive, owing no doubt to the reflectivity of that mass of water. The color on the horizon on a clear day on the Keweenaw Peninsula, surrounded by water, particularly noticeable in the fall, is luminous and falsely tropical. This peculiar quality of light and color is yet another reason why it is impossible to forget the presence and importance of the lake in making the landscape. There is little wonder that "the savages revere this Lake as a Divinity, and offer it Sacrifices, whether on account of its size,—for its length is two hundred leagues, and its greatest width eighty,—or because of its goodness in furnishing fish for the sustenance of all these tribes, in default of game, which is scarce in the neighborhood" (Thwaites 1902:31).

The discipline of anthropology explains that nature and human culture are everywhere tightly associated and in fact do act on one another as inseparable partners. This tenet is easy to appreciate in an environment as variable and dynamic as that of the Lake Superior basin. The lake's size (ca. 600 km from east to west and ca. 200 km north to south) has a profound effect on local climatic conditions. The climate and weather within about 30 km of the lake shore is a product of seasonal shifts in the temperature differential between the lake and its adjacent landmasses, and local microclimatic conditions are a function of distance from the lakeshore. In the winter this means that overcast, severe, and seemingly endless snowfall and relatively cold temperatures can be expected in the highlands. Absolute near-shore lands receive less snow, as does the interior 100 km to the south. But the Keweenaw Highland receives, on an average, greater than 5 m of snow per year, and the snow cover may last in excess of 140 days (Sommers 1977:51). In peak years the snowfall can exceed 10 m. Extremely long, cold winters quickly followed by compacted springs, summers, and falls are reliable seasonal patterns.

> Frosts, of sufficient severity to turn the leaves, usually occur as early as the middle of September. Snow commences falling by the middle of October, and for more than six months the ground is covered with a

fleecy mantle. The streams become locked with ice and remain so until May. . . . Spring and summer are mingled. The forest become clothed with leaves, and its solitude is enlivened by the song of birds and the hum of insects, before all traces of snow have disappeared. (Foster and Whitney 1850:56)

Mean temperatures in the trap range average 14–16 F in the winter, and ca. 66 F in the summer, but there is significant microvariation depending upon local setting and wind direction. Lakeshore days and nights are commonly twenty degrees F warmer than in the interior during the winter. In the summertime, fogs and cooler temperatures than those in the interior prevail at the lakeshore. The lakeshore areas are particularly susceptible to sudden changes in weather, and both winter and summer storms can be extremely violent.[4] Schoolcraft's remarks upon completing a canoe trip, from Sault Sainte Marie to the Ontonagon River along the south shore of the lake, stress the abruptness with which conditions changed on the lake. "Gusts of wind, arising with a momentary warning, have often driven us hastily ashore; and the whole route may be characterized as stormy, and yet we are told this is one of the most favorable months for performing the journey. In the autumn it is seldom attempted. The winds, which generally prevail from the northwest, expose the southern shore to the fury of continual storms" (1970:168–69). Foster and Whitney found evidence of their severity on the mainland: "Thunder-storms of great violence are not unusual; and the large tracts of prostrate timber frequently met with in the forests, and known as 'windfalls,' indicate the path of the tornado" (1850:57).

Abrupt transformations in conditions characterize the local environments of the trap range, the Keweenaw Peninsula, and the Lake Superior basin as a whole. This is particularly the case with spectacular meteorological events. Violent hurricane-force windstorms strike the land from the water bodies, creating seiches, waterspouts, and phenomenal cloud formations. Clear nights display auroras of movement and color variation, filling two to three hundred degrees of the sky with displays of the northern lights. Foster and Whitney painted the aurora phenomena this way. "The commonest phenomena are these: A dark cloud, tinged on the upper edge with a pale luminous haze, skirts the northern horizon. From this, streaks of orange and blue colored light flash up, and often reach a point south of the zenith. They rapidly increase and decrease, giving to the whole hemisphere the appearance of luminous waves, and occasionally forming perfect coronae" (1850:57).

Another of the astonishing kinds of meteorological transformations over Lake Superior's horizon is a peculiar mirage phenomenon, an optical illusion resulting from severe differentials in air and water temperatures and in the fluctuations of air layers of varying temperatures. They are so extreme as to be disorienting, and they occur more frequently and with greater effect over Lake Superior than over any other of the Great Lakes. Because they are products of air movement and differential temperature, the mirages are dynamic. They rapidly change shape and position; they appear to hang in the sky or float above the water's surface. Typical mirage phenomena include distorted reflections of landscape features and terrain, which may appear as disproportionate doubled images, or even as inversions of actual shapes. Foster and Whitney described these phenomena in detail:

> Mountains are seen with inverted cones; headlands project from the shore where none exist; islands, clothed with verdure or girt with cliffs, rise up from the bosom of the lake, remain a while and disappear. In approaching Keweenaw Point, Mount Houghton is the first object to greet the eye of the mariner. Its dome-shaped summit serves as a landmark to guide him in his course. Once or twice, in peculiar stages of the atmosphere, we have observed its summit inverted in the sky long before the mountain itself was visible—Thunder cape would assume shapes equally grotesque: at one time resembling a huge anvil with its handle projecting over the lake, at another it appeared as though transversed from summit to base by an immense fissure. (55–57)

These transformations are, to some degree, repeated as themes in the beliefs of the native people of the region, at least as reported in historical narratives about the Lake Superior basin (Foster and Whitney 1850:10). In fact there is nothing as likely as transformations of state according to many of the traditional stories of native belief. The aboriginal people, of course, had equally compelling explanations of the causes of these fluid transformations. It was the culture hero Nanabozho, or his enemies the Mishi Ginabig. Or perhaps it was Mishi Bizi the water panther, wielder of power over water and the underground, who was operating in and on the material world, transforming it to his momentary and potentially damaging whimsy. Appreciate that these spectacular changing clouds and mirages actually behave in ways that are perfectly logical in local native cosmology. In fact this feature of distortion is a theme that is repeated in beliefs about other transformations of state. What could be more likely? All of these images find their way into descriptions of nature as seen by the region's native people; they are considered to be caused by cosmic events and actions of powerful person-forces, or manitous.

Conclusions

The western Lake Superior basin is a unique environment. Its most distinctive qualities derive from its ancient copper-bearing bedrock and the massive scale of the lake itself. Glaciation radically altered this environment, made the present topography such as the lake basin and its shorelines, conditioned local microenvironments to a new mix of floral and faunal communities, exposed copper-bearing bedrock, and redistributed copper all over the Midwest. The cessation of ice cover preceded the complete revision of local ecological relations; humans soon added themselves to the system and the landscape was remade.

Careful comparisons and the support of ethnographic, historical, and archaeological data enabled archaeologists to reconstruct probable patterns of past resource use. The most important prehistoric Lake Superior resources in terms of reliable food were fishes such as trout, whitefish and sturgeon, maple sap, wild rice, berries, and tubers; large mammals such as deer, bear, and moose; and fur bearers such as beaver. Important nonfood resources were birch bark, basswood, and copper.

The Lake Superior environment is unusual in the continental United States because of the extension of subarctic conditions to southerly extremes; this is particularly true of Isle Royale: "its subarctic flora and fauna; its clear, cold waters; its glorious effects of light and shade, sunshine and shadow; its strange mirages; its mysterious tides; its beautiful gems; its everlasting rocks; its wild scenery, pictured headlands and isles; and its evidences of the Indian miners" (Dustin 1957: 33–34). This flowery description fits the coastline of the western basin most aptly. A cultural landscape is both a composite record of human activities and a set of perceptions about the causes of natural phenomena. There is strong correspondence between natural phenomena such as lake-induced mirage and the northern lights, and the stories that native people told about supernatural creatures that inhabited the region and their antics. These stories center on Lake Superior and are congruent with its natural history.

Suggestions for Further Reading

Bornhorst, Theodore J., and William I. Rose. *Self-Guided Geological Field Trip to the Keweenaw Peninsula, Michigan. Proceedings of the Institute on Lake Superior Geology* 40, no. 2 (1994). An account of the prominent features and patterns of bedrock and glacial geology, planned as a set of car trips and hikes, including dozens of detailed maps.

Foster, John W., and Josiah D. Whitney. *Report on the Geology and Topography of a Portion of the Lake Superior Land District in the State of Michigan. Part 1: Copper Lands.* House of Representatives, 31st Cong., 1st sess., Doc. no. 69, Washington, D.C., 1850. Details the early discoveries of prehistoric mining evidence in the Lake Superior region and provides a thorough environmental description and an account of early industrial mining companies.

LaBerge, Gene L. *Geology of the Lake Superior Region.* Phoenix: Geoscience Press, 1994. A description of the bedrock and glacial geology of the region, written for the nonspecialist; profusely illustrated.

3

What Has Particularly Recommended Their Labors to Me Is Their Love of Truth (Gallatin to Henry, June 12, 1847)

—Ephraim G. Squier and Edwin H. Davis, 1848

Outstanding Personalities

> It has been a constant aim in the preparation of this memoir, to present facts in a clear and concise form, with such simple deductions and generalizations alone, as may follow from their careful considerations.
> —Ephraim G. Squier and Edwin H. Davis, 1848

Introduction

Over the past one hundred and fifty years the many students of America's early history assembled the story of prehistoric copper working. Their research began about 1845, and its reach extended not only across northern Michigan but to the central rivers of North America and as far afield as the British Museum. The co-authors of this story were natural scientists, geologists, journalists, metallurgists, lawyers, captains of industry, students, lay persons, scholars, and the idle rich. The following narrative gives these discoverers faces and personalities and reveals how they came to the conclusions that they did. In some cases, the stories of these people's exploits were as interesting as their archaeological conclusions.

Not all of the many accounts that exist about the earliest discoveries of the prehistoric mining evidence are included here. They are not all equally accurate or original. This narrative features the writings and accounts of people whose firsthand observations, inferences, and deductions are methodologically valid. That does not mean that they came to the same conclusions as one might in the present day. It simply means that their conclusions were well reasoned: their assumptions were clearly identified, their evidence carefully described, and their chains of logic meticulously linked. They were people who, as archaeologists everywhere should, depended primarily on material evidence to tell the stories of the past. Likewise they were people whose findings reflected lines of inquiry, ways of thinking, and modes of questioning that were particularly archaeological in character.

The title of this chapter refers to the special qualities these researchers pursued as they did their work. What marks the work of these writers is the respect they share for common standards of evidence. It would be too great a gloss on history to suggest that they never erred, but then the same could be said of all great scientists regardless of discipline. The way of science is littered with wrecked theories, discarded papers purporting to represent truth, and disproved conclusions. It should be that way because science, after all, is based on error recognition. The goal of science is, in part, to disprove what has already been learned and accepted.

Despite this praise, the basic human characteristics of our scientists are important to consider. They were all prone to assumption and errors of judgment, just as nonscientists are. All thinkers are somewhat bound by circumstance, which is the result of cultural experience and its presumption of and encouragement of predictable habits of thinking and behaving. Because these people were creatures of their own cultures, they tended to adopt the cultural beliefs around them as accepted truths. To some degree they may have been unconscious of their own cultural biases. For example, it was unlikely that the average nineteenth-century scientist could put aside assumptions about the technological importance of metals. After all, these scientists were the issue of a culture that prided itself in applying modern metallurgy, and its contemporary applications had expanded beyond their wildest imaginations. Universal notions of progress, in which the unbreakable march upward in technological achievement apparently led to the sky and beyond, had been a part of Western thinking for thousands of years. These pet assumptions were not abandoned by the scientists of the last century, nor of this one. But some thinkers were more willing and able than others to let such assumptions rest quietly and to lay them aside momentarily as they considered novel orientations to evidence. These are the people who are included here.

Squier and Davis Discover Metals in Context

The team efforts of Ephraim Squier and his co-worker Edwin Davis focused on the curious prehistoric mounds that occupied the river terraces of the central United States. The mounds sometimes contained copper artifacts, and this was considered a remarkable achievement of the ancients who made the mounds. The copper artistry was unexpectedly sophisticated in the eyes of nineteenth-century Euro-Americans. Squier and Davis spent the balance of the late 1840s mapping, excavating, collecting, and analyzing the mounds, and their contents. Their results were published as the very first number of the series *Smithsonian Contributions to Knowledge* (Squier and Davis 1848). The series was to be multidisciplinary and the instrument to carry out James Smithson's primary intent in bequeathing his estate to the United States: "To carry into effect the purposes of the testator, the plan of organization should evidently embrace two objects,—one, the increase of knowledge by the addition of new truths to the existing stock; the other, the diffusion of knowledge thus increased, among men" (iii).

There were, however, some caveats regarding the acceptance of manuscripts and content. They were to be blind-reviewed carefully and, most important, "no memoir, on subjects of physical science, [was] to be accepted for publication, which does not furnish a positive addition to human knowledge, resting on original research; and *all unverified speculations to be rejected*" (iii; emphasis added). Believing that their work fit this bill fully, Squier and Davis submitted their manuscript for peer review on May 15, 1847. It was dutifully pored over by a committee of members of the American Ethnological Society, who enthusiastically recommended it for publication.[1]

As skeptical scientists, Squier and Davis were extremists by today's standards. They believed that it was not only possible but necessary to put aside all presumptions, biases, speculations, and favorite theories as one examined evidence. And this they set out to do. "At the outset, as indispensable to independent judgment, all preconceived notions were abandoned, and the work of research commenced *de novo,* as if nothing had been known or said concerning the remains to which attention was directed" (xxxiv). The Squier and Davis view of appropriate scientific procedures (enforced by the editorial hand of Joseph Henry, secretary of the Smithsonian) assumed that it was possible to begin theoretically afresh and that they could rid themselves of any opinions that might interfere with pure deduction. The outcome they sought was not only clarity, but systematic investigation and rapt attention to every detail of method and procedure. To them, science and speculation were intrinsically different and required separation. Their intentions were very much in anticipation of anthropological theories of a half century later, when it was widely believed that facts were finite, capturable, and equally accessible regardless of the theoretical orientation of the researcher. In fact, Squier and Davis claimed to proceed without hypotheses at all and professed equal skepticism of arguments by analogy: "With no hypothesis to combat or sustain, and with a desire only to arrive at truth, whatever its bearings upon received theories and current prejudices, everything like mere speculation has been avoided. Analogies, apparently capable of reflecting light upon many important questions connected with an enlarged view of the subject, have seldom been more than indicated" (xxxviii).

They then identified those queries that they believed to be unanswerable. *Their data, they concluded, were inadequate to explain the relations, if any, among American sites and those of the rest of the world.* They also confessed a shortage of details and inadequate evidence on mound construction details, contemporaneity of mounds, and evidence for specific systemic ties among sites. Finally they drew a remarkably modern conclusion about the future of the mounds. The monuments were about to be destroyed by the onslaughts of human recolonization and development. They identified precisely the same agencies of destruction then that we know are commonly responsible for site ruin now. Their solutions presaged those of the late twentieth century: local-level responsibility and government stewardship.

> The importance of a complete and speedy examination of the whole field, cannot be over-estimated. The operations of the elements, the shifting channels of the streams, the levelling hand of public improvement, and most efficient of all, the slow but constant encroachments of agriculture, are fast destroying these monuments of ancient labor, breaking in upon their symmetry and obliterating their outlines. Thousands have already disappeared, or retain but slight and doubtful traces of their former proportions. Such an examination is, however, too great an undertaking for private enterprise to attempt. It must be left to local explorers, to learned associations, or to the Government. (xxxix)

As did other scientists of their age, Squier and Davis suffered from a rudimentary view of time-space systematics.[2] They interpreted the eastern prehistoric mounds as part of one large system which was superseded by the cultural systems of the native peoples encountered by the earliest invading Europeans.[3] But they were obviously aware that there was a great deal of variation in size, function, density, and arrangement of the mounds. They clearly recognized the association of the mounds with major river valleys and promontories of geography, including the careful placement of sites at terraces as well as defenses on hilltops. They also noted that the mounds were most often situated in places considered the most favorable for future settlement by Euro-Americans: "The centres of population are now, where they were at the period when the mysterious race of the mounds flourished" (1848:7). They identified the Effigy Mounds of Wisconsin as culturally distinctive; likewise the garden beds of Michigan and the larger mounds and earthworks of Ohio and the south appeared to be culturally separate.

Deliberations about the ages of the mounds were too speculative to be entertained. Resting squarely on geomorphology and dendrochronology, however, the researchers did ascribe mound localities on second river terraces to relatively older times. In addition, tree growth near the mounds was essentially no different than the composition of forests outside of the mound distribution area. Squier and Davis concluded that full forest regeneration, something that must have taken centuries, had occurred.[4] Overall, they treated questions of relative antiquity as gingerly as possible. "It is not undertaken to assign a period for the assimilation here indicated to take place. It must unquestionably, however, be measured by centuries" (306).

As emergent archaeological practitioners—there was no such formalized discipline at that time in the United States—the two were adamant about the critical importance of the Law of Superposition.[5] The two researchers required that things unequivocally associated with the moundbuilders by necessity be found under the mounds. Some of the copper axes and other copper implements were not found that way, so to Squier and Davis their association with the builders was problematic: "It will be seen that they are not found where, as a general and almost invariable rule, we must look for the *only authentic* remains of the mound builders, viz. at the bottom of the mound" (198). Thus for Squier and Davis the position of artifacts relative to one another and to humanly created features was the key to understanding and authenticating their associations.

Some of Squier and Davis's reasoning about artifact functions appeared to derive from experimental replication of objects that they discovered within the mounds. They identified drills or gravers as sculpting tools for stoneworking and appeared to have experimented with them to demonstrate this possibility: "It cuts the softer varieties of stone with facility" (200). They suggested multiple functions for awls and gravers, including their possible function as bars for bending and subsequent use as ornaments. In this sense they were very modern in outlook, adopting the view that context revealed meaning rather than holding to the view that objects had singular invariant functions, as the functional labels "awl" or "graver" might imply.

The work of Squier and Davis was absolutely contemporary with the first systematic descriptions of prehistoric copper working from the Lake Superior district. They were immediately aware of the importance of the discoveries on the Michigan Copper Range, as well as conversant with what they believed were the unique characteristics of Lake Superior copper. An examination of how Squier and Davis incorporated the subject of copper working into their volume reveals much of their scrupulous methodological orientation and their interest in solving archaeological dilemmas. Metalworking was the primary subject of no fewer than three chapters; it was doubtless a subject that created great controversy and speculation prior to their work. For example, it was widely reported about 1848, gold fever being on people's minds, that gold had been found in mounds. These two simply asked for independent verification of the reports of gold. When it did not exist to their satisfaction, they actively dismissed such reports as hearsay, not evidence. Two remaining questions were of particular interest to them. First, they considered where the sources of the raw copper were to be found. Their approach was based upon a comparative methodology and ancillary data from the discipline of systematic geology. Visible silver was the key to their sourcing argument. They reiterated and expanded their arguments, being well aware of the discoveries taking place in the Lake Superior copper district even as they wrote:

> In consideration of the amount of the metal discovered, implying a large original supply, and the fact that it is occasionally found combined with silver in the peculiar manner characterizing the native deposits upon the shores of Lake Superior, we are led to conclude that it was principally, if not wholly, derived from that region. This conclusion is sustained by recent investigations upon the shores of that lake. These have led to the discovery that the aborigines, from a very remote period, resorted there to obtain the metal. There is also evidence that some of the more productive veins were anciently worked to a considerable extent. (279)

Second, they addressed speculations that the copper-working technologies were evidence of precolonial European contact. Squier and Davis's reasoning deserves careful attention. The question was a well-developed one among dabblers in American prehistory, and it was not conclusively resolved by Squier and Davis nor by others until more than a century later.[6] Squier and Davis addressed four lines of evidence: association with objects of unassailable European production; levels of metallurgical technique; geographical sources of copper; and arguments based in an understanding of depositional processes. "With respect to the question whether these remains are of European origin or manufacture, I have merely to remark that their workmanship is very rude; that no traces of iron or of European implements were found with them, and that the copper corresponds exactly with the specimens of native metal obtained from Lake Superior" (202). The depositional process arguments were essentially also arguments of association. They stressed that the level of bone preservation of accompanying skeletal remains and the differences in both burial pattern and placement from those of the historic Indians suggested that "the antiquity of the materials seems assured" (202).

Though Squier and Davis were unprepared to offer detailed solutions to questions of moundbuilder contacts with cultures from the rest of the prehistoric world, they instructed that exotic (nonlocal) materials such as copper and obsidian were the best sources of understanding patterns about migration, origins and interactions within moundbuilders' cultures. "The discovery of *obsidian,* a purely volcanic production, in the mounds, in a region entirely destitute of the evidences of immediate volcanic action, is, to the commonest apprehension, a remarkable fact, a subject of wonder; but neither marvels nor mysticism have aught to do with science" (278). They suggested that speculations were pointless on the question of migrations; the materials, after all, were eventually able to tell the tale. They concluded that the existence of multiple cultures across geographic extremes was well evidenced by the law of association of objects; things from all over were deposited together and at the same time. To them, this fact argued against migration theories, which sought to derive full-blown mound cultures from elsewhere in the world.

Overall, Squier and Davis applied emerging archaeological principles of reasoning to questions presented by the mounds and their associated materials. They repeatedly rejected speculation in favor of simpler explanations, which were congruent with natural depositional processes, known geographical distributions of materials, and observable processes such as patterns of forest succession. Over and over again their evidence was material, and its associations and contexts were the keys to understanding the age and origins of American antiquities.

Foster and Whitney Apply a Parsimonious Perspective

Meanwhile, somewhat to the north of Ohio, important discoveries occurred on the shores of Lake Superior. Just as Squier and Davis documented the mounds of the central United States, a team of their contemporaries, the geologists John W. Foster and Josiah D. Whitney, were recording the distribution and contents of the mineral-bearing lands of the lake district (figs. 6, 7). In six brief pages they described the discoveries of prehistoric mining made three years earlier on the Ontonagon portion of the trap range (Foster and Whitney 1850:158–63).

This reportage, so close to real time in appearance, drew its content from firsthand observation, as well as the scrupulous methodological procedures of its authors. As geologists Foster and Whitney possessed, first and foremost, a material and spatial orientation to bodies of data. Their overall approach had much in common with developing archaeological practice. Most importantly, they based their assumptions and conclusions about early mining on direct observation of material evidence. This evidence included hammerstones, scarred bedrock, tailings piles, shovels and ladders, ash beds, bailing equipment, and the distribution of mining pits in relation to the

Fig. 6. John W. Foster. (Photo from Merrill 1924:280, courtesy of Yale University Press.)

Fig. 7. Josiah D. Whitney. (Photo by Silas Selleck, San Francisco, 1863. The California Faces Collection, Item 5; courtesy of the Bancroft Library, University of California, Berkeley.)

copper veins. They linked the distribution of the mining evidence from the Ontonagon district with that of the Keweenaw Peninsula and Isle Royale. By methods of comparison and techniques of measurement they described the variability within the samples of hammerstones that they examined and illustrated what they had discovered. They drew argumentative analogies with ancient mining techniques in other parts of the world, most notably the Harz Mountains of Germany. And they deplored the failures of their contemporaries to preserve the ancient evidence, taking Mr. Samuel Knapp of the Minesota Mine, the actual discoverer of ancient mining, to task in particular, chiding that he showed "little reverence for the past" in his use of ancient hammerstones to wall up a troublesome spring (160).[7]

What motivated the original miners to seek copper was, to Foster and Whitney, a simple matter of recognizing the special qualities of the metal as a material for tool making. "A race like the Indians, dependent principally on hunting and fishing for the means of subsistence, would employ copper, where it was accessible, in the construction of their weapons of capture, in preference to stone, it being more easily fashioned and less destructible. This would naturally be expected in the rudest and most simple state of society" (162).

As did Squier and Davis, the geologists addressed the perennial questions of the identities of the miners and the antiquity of the ancient mining with great caution. Their scientific conclusions were parsimonious; that is, the simplest explanations were assumed to be the most likely. Their inferences from dendrochronology, gathered by counting tree rings from mature trees growing from mining pits, suggested to them that the mining took place as early as A.D. 1400 (160). Their assumptions from the start were that there were two likely candidate groups for the miners: the precursors to the local native populations, or the moundbuilders. They found the notion of an ancient "race" of miners to be unlikely. "All will admit that the facts above set forth assign to these excavations a high antiquity; but whether they were made by a race distinct from the Indians now inhabiting the region, is a matter of extreme doubt, although all traditions with regard to their origin have vanished" (162). The reason that this "other race" explanation could be demoted to doubtful was perfectly simple; there was no evidence. A report of mysterious standing stones, plow scars, and what were called "Tartar inscriptions" was dismissed for the same reason (163).

Foster and Whitney interpreted the absence of local native knowledge about mining as evidence for its deep antiquity. Their logic around this question was quite clear; they suggested that mining ceased when novel materials became widely available through trade with incoming Europeans. In other words, they suggested that Indian ways of life changed during contact with Europeans, hardly a mysterious prospect but one evidenced widely across the continent. They suggested that historical research into the region's earliest narratives might reveal details about copper technology prior to their disruption by Europeans and would likely provide a direct historical link with the present.

In considering the strengths of the Foster and Whitney chronicles and the validity of the conclusions to which they came, their essential qualities should now be evident. They stressed parsimony of explanation, reliance on material evidence, explicit identity of assumptions, careful use of comparison and analogy, and suggested promising further research.

Wood and Wilson Are Eyewitnesses of Some Repute But Make Some Errors in Observations

Both Daniel Wilson and Alvinus Brown Wood reported on the appearance of the mining evidence shortly after the first descriptions were published. Both visited the mainland mining districts in 1855, and Wood was likely there even earlier. Wilson (fig. 8), a Scot-Canadian and professor of literature and history, toured the discoveries in Ontonagon, and his accounts were of interest for a number of reasons. He applied and refined the comparative method, that of drawing analogies between the mining evidence in North America and remnants of ancient mining cultures elsewhere. Viewing the ubiquitous hammerstones of Ontonagon, he began with a simple observation: "I was greatly struck with the close resemblance traceable between these rude mauls of the ancient miners of Ontonagon and some which I have seen obtained from ancient copper workings discovered in North Wales" (Wilson 1856:228). The cause of observed similarities was then pursued. Rather than attributing the gross similarities of form and function to Welsh miners invading the Superior basin, Wilson reasoned that cultural contact was an unnecessarily complicated cause for the fact that stone hammers tended to look alike. Instead, he suggested that people confronting similar circumstances tended to solve similar problems in similar ways (228).

In keeping with other observers of his age, Wilson never comfortably resolved the question of the link between moundbuilders and the miners; he favored a speculation that an invasion of what he called "Red" Indians, or perhaps pestilence on a scale comparable to that caused by the European entrada, put a stop to both mining and moundbuilding. This idea was in part based on his erroneous assumptions that the miners had suddenly and irrevocably stopped their work, and he may have been the first to advance this thesis in print. "Certain it is that the works have been abandoned, leaving the quarried metal, the laboriously wrought hammers, and the ingenious copper tools, just as they may have been left when the shadows of the evening told their long-forgotten owners that the labours of the day were at an end, but for which they never returned" (237).

Wilson's final commentary, based on observation rather than speculation, was that the prehistoric copper tools had not, contrary to popular notions, been hardened by some mystical metallurgical practice, but by the simple process of hammering, which he pointed out was a fact known to all metalsmiths. In addition, he suggested that the native practice of copper working, reduced somewhat in scale and likely in frequency, existed at the time of Alexander Henry and other chroniclers of the old Northwest, who reported widely on its aboriginal use as ornamentation and its value in trade (237). His accounts were, in summary, able to clarify some misinterpretations, but fell into fits of speculation, creating others.

This is also true of his contemporary, Alvinus B. Wood, who in the early 1850s observed the opening and clearing of a number of prehistoric mining pits. In 1906 he reported to the Institute of Mining Engineers an

Fig. 8. Daniel Wilson. (Photo from Appleton's *Cyclopedia* 6:574, courtesy of the MTU Archives and Copper Country Historical Collections, Michigan Technological University.)

eyewitness version of the mining evidence. Wood was important because of his early firsthand accounts, his proneness to the experimental mode, and his knowledge of the range of copper's utility. He reported on the dimensions and character of the mining pits at the original Quincy Mining Company main shaft and plotted the distribution of mining pits elsewhere across the Keweenaw district on both sides of the Portage Canal, citing those industrial age mines that had begun on ancient diggings. In order to understand the amount of copper potentially available to the ancients, he proceeded in an experimental frame: "To get an idea of the amount of copper which the ancients obtained, I had the uncovered surface—30 by 16 ft—worked over with pick and bar, no powder being used. Ten tons of barrel-copper were taken up; the largest piece weighing 1,500 lb" (Wood 1907:291).

Wood's description of the utility of some forms of native copper over others made it clear that he understood some of the limits of prehistoric metalworking technology. Medium-sized pieces had to have been the most useful, and those too little and too large were left behind (289). Some deposits yielded better materials than others, given prehistoric preferences. "The ancient miners worked only where the "wash" covering the cropping of lodes or veins was thin, and the lodes were easily found" (290). To Wood, it was apparent, on observing this regularity in undisturbed settings, that deeply buried lodes were not explored. Rather, he commented that in one place, the mining was done on a surface having "the benefit of disintegration since the Ice Age" (291). And in another place, "nearly north of the North West mine, a number of ancient pits were found on a small fissure-vein, which carried sheet-copper well suited to the uses of the old miners." Likewise on Isle Royale, "extensive ancient mining was done on the easterly part of Isle Royale, on narrow veins yielding sheet-copper. This form being, as already observed, suited to the needs of the old miners, they did much work on these veins" (292).

Unlike virtually every other commentator of the time, Wood was not particularly impressed with the scale of ancient mining, though he was very interested in modest calculations of their total take. "The copper they obtained by working the ancient mines, as shown by the part estimates on previous pages, must have amounted to several hundred tons. They also had the first picking of the float-copper, dislodged from the lodes and veins during the Ice Age, which must have amounted to many tons. The present inhabitants have picked up many thousands of pounds" (294). In fact he wondered why they had not done more mining than they did. Regrettably, Wood fell hard for the prevailing belief that the miners were in fact migrants whose real homes were elsewhere. For Wood, the miners' homeland was Mexico, for he identified them as Toltecs or perhaps Aztecs (296). To Wood, the evidence of their short-lived occupancy lay in the partially burned wood in the pits, which he attributed to the miners' rapid departure.

Wood's direct observations, like those of Wilson's, cleared up some mysteries but prolonged others. Credit is due to both observers for their experimental approaches, their knowledge of copper as a malleable material, their ability to draw on comparative bodies of data, and their ability to connect the aboriginal use of copper to the historic record. The shortcoming they shared was the attribution of ancient mining to short-term users who departed the area with abruptness. This belief would last for several more decades.

Charles Whittlesey Offers Contradictory Scenarios

Friends with Squier and Davis, likely crossing the paths of Foster and Whitney, Charles Whittlesey combined his knowledge of archaeology and geology to produce two seminal reports on prehistoric technology (fig. 9). He published in the *Smithsonian Contributions to Knowledge* series, contributing to volume 3 in 1850 and volume 13 in 1863. The 1850 volume consisted of maps and plans of Ohio mounds that he assembled in 1838–39 as a member of the Ohio Geological Corps. These may have been reviewed by Squier and Davis, but they were not published, and Whittlesey offered them as addenda to the first volume. He was rather at odds with Squier and

Fig. 9. Charles C. Whittlesey. (Photo from Merrill 1924:452, courtesy of Yale University Press.)

Davis's conclusions about the mound origins, however. "I do not feel inclined to attribute the great works of Central and Southern Ohio to the progenitors of our Aborigines; but in regard to those of Wisconsin and Minnesota there is room for doubts and ample discussion on this point" (Whittlesey 1850:6). This opinion would skew his conclusions about ancient mining on the Lake Superior district.

It was number 155 of the *Smithsonian Contributions to Knowledge,* which appeared in volume 13, that reported on Whittlesey's firsthand experiences in the copper lands. His paper, "Ancient Mining on the Shores of Lake Superior," was accepted for publication in April 1862. The data it depended upon were gathered by Whittlesey in 1851 on the south shore of Portage Lake, Houghton County, Michigan, and probably from Isle Royale.

There was a great deal of unevenness in his presentation of his findings; depending upon the topic at hand his adherence to scientific skepticism varied. His remarks were masterpieces of an extreme skeptical approach, as long as his topic was geological processes. Mindful about the great debate which was then raging about the origins of glacial drift, he did not credit Agassiz' theories of glacial ice sheets, preferring the relative anonymity of "that agent, whatever it was" for the deposition of drift (2). In 1851, explanations of continental glaciations were not agreed upon, and Whittlesey took no chances in championing one or another specific drift depositional mechanism. He was willing to suggest that the further from the bedded source, the smaller the pieces of copper in the drift tended to be (2).

In contrast, when it came to human cultural behavior, Whittlesey freely and rather carelessly assumed to understand fully the thinking, motivations, and ideologies of the known historic inhabitants of the region. "As a people, if we may judge by their silence on a subject on which they may be supposed inclined to be communicative, if they had anything to tell, the aborigines have no traditional knowledge of their predecessors, the race of the 'mound builders' " (3). But when it came to attributions of great skill to the ancient miners, Whittlesey was second to none in his appreciation of them. This set of attitudes was unfortunately consistent with his reluctance to see nineteenth-century native people as the direct descendants of the early miners.

One of Whittlesey's primary concerns was to sort the historical record into two piles, one composed of people who saw things themselves, and the other of those who said they did. His approach was one of purest skepticism about the stories he was told about aboriginal people; however, he himself was willing to ascribe firm motivations to them. His questions revolved, again, around the themes of miner identity and relative antiquity. Whittlesey intended to show that at the beginning of the contact era Indians did not actively mine, "but only rudely fashioned copper knives, that were evidently beaten out from small boulders. . . . They had no implements proper for the purpose; nor did they produce samples of metal taken from its position *in situ*" (2). Historic Indians had other interests. "Instead of viewing copper as an object of everyday use, they regarded it as a sacred Manitou, and carefully preserved pieces of it wrapped up in skin in their lodges for many years; and this custom has been continued to modern times" (2–3).

Then he suggested that the recent people inherited no knowledge of mining from their forebears. According to Whittlesey's logic, if they had, they would have told their Jesuit spiritual fathers at length about such knowledge. He reasoned that the missionaries did not record minute details about mines, because they never came across any such information. "If the Indians possessed traditions from their ancestors relating to ancient mines, or the people who worked them, those must also have come to the ears of the Jesuits" (3). He concluded that the mining activities must have been from a remote time and that other explanations made little sense given tradition and necessity. Thus he discounted the attribution of mining to the native people of the region.

Whittlesey was, however, a meticulous observer of material detail, which made his reports of great use to other researchers. He concluded that mining pits were difficult to distinguish from ordinary uneven ground. "They are, for the most part, merely irregular depressions in the soil, trenches, pits, and cavities; sometimes not exceeding one foot in depth, and a few feet in diameter" (5). Their antiquity was supported by the accumulation of debris within them, as well as a layer of reddish sediment, which he claimed indicated "the long period of time since the excavation was made" (5). He was meticulous in describing the geological strata revealed at each mine, the prominent topographic features, and the material culture and other organic remains lying about the sites. He illustrated carefully drawn profiles of stratigraphy and included detailed accounts of use-wear patterns on prehistoric tools such as shovels and bowls (7–8). He made careful observation of shatter marks and pick marks on trap rock wall faces and compared them with those made by European tools.

Whittlesey noticed characteristics of the materials recovered as well as their potential functions. For example, he identified that hammers with no groove were only battered at one end, and reasoned that they were used by hand "as though the manner of holding them was such that a blow was not given on the other" (12). He also assumed that there was a way of hafting smooth stones and that people had their own peculiar favorite ways of using stones. "Different parties of men may have preferred tools of different kinds, which would account for mauls, which are seen at one mine, being among themselves alike, but dissimilar to those at other places" (13).

But some of this commentary was not solely based on empirical data. For example, his suggestion about the relative ease of melting copper was inconclusively linked with firsthand experience. "It is quite singular that they had not discovered the art of melting copper, which can be effected so easily in an open fire made of wood, but no evidences have fallen under our notice that this was done by that ancient race" (13). Many others have demonstrated that it is not a simple matter to heat copper to its melting point at 1083 C; most experimentation suggested that charcoal and/or a sufficient airstream were necessary to reach that temperature. Here Whittlesey appeared to be on his careless side.

In contrast, his comments on ancient excavations in the drift by Quincy landing at Portage Lake in Hancock appeared to be based on firsthand observation. Here, Quincy Mine officials claimed that copper was such a common constituent of the glacial drift that the company thought it would pay to water wash the gravel (15). Whittlesey dutifully recorded the presence of ancient pits in the gravels and remarked that this was also a convenient place to find useful mauls of a proper size and shape.

Making notice of the patterns of copper extraction practiced by the ancient miners compelled Whittlesey to praise their abilities in predicting copper's courses. "On this elevated ground the old operators had discovered and worked a rich deposit of copper which was nowhere visible upon the surface. The direction of the line of pits is northeast and southwest, corresponding with the range" (16). Here credit was given to the old operators for discovering that the Portage Lake volcanics were true lodes and ran parallel to the peninsula, a fact which had little to do with their visibility on the surface. In other words, once this was known, their paths were predictable.

Pursuant to understanding more about prehistoric mining, Whittlesey undertook exploratory excavations in 1851 to ascertain whether patterns of pits lying to the south of Portage Lake were indeed ancient mining evidence. This was necessary because the lay of the trap rock on the south of the Portage differed considerably from that to the north. And he was skeptical about this undertaking. "It required some assistance of the imagination to conceive that the slight and irregular depressions, which were dimly visible among the trees, were the works of men" (16). But in trenching across the ancient excavations to prove their origins, he discovered that the mining pits matched those of the bearing of the formation. Further to the south at Ontonagon, the formations ran with the range and not across it as they did on the far north Keweenaw; this the old operators also knew (17).

Whittlesey apparently examined some of Samuel O. Knapp's original discoveries of prehistoric mining at the Minesota Mine diggings firsthand, for he was on the Ontonagon district in 1849 (18). He claimed that he took prehistoric wood from that mine, specifically pieces of oak that underlay the famous mass originally discovered there and that had been identified as skids, with marks of a cutting tool very clearly in evidence. Whittlesey examined maul scars both on boulders and on the walls of the trap rock, and he also noticed evidence of fire within. "The marks of fire on the rocks of the walls are still evident. Charcoal, ashes, and stone mauls are found in all of the pits hitherto cleaned out" (18). He personally counted the rings on a tree growing out of the pit where the original worked mass was found. Whittlesey pushed the apparent age of the mining back even further by his notation that "fallen and decayed trunks of trees of a previous generation were seen lying across the pits" (19).

He found the skill with which the ancients excavated useful copper to be remarkable. They created artificial cavities in the "mural faces of trap at various elevations and wrought upon it extensively, seeking with assiduity for the rich portions, no matter how difficult of access. Some of their excavations on the side of the bluff are scarcely large enough to shelter a bear" (20).

Whittlesey stated his reasoning for ascertaining that the mining was not done by the "present Indian race" (26). This section was again an exhortation supported by negative evidence, in which he freely interpreted the motivations and capacities of the miners. According to Whittlesey's unbridled speculations, the miners had "no domicils, graves, cities, highways." Notwithstanding this, they appeared to him to be technologically more advanced than nineteenth-century Indians. He surmised that they had better transport and bigger boats, in which they carried needed supplies for several months, leaving before winter for home with a load of copper. They took the bodies of their dead home with them, and few females and children made the trip anyway. They used mostly Lake Superior copper in the mounds, and the Ohio chisels reported by Squier were the kind that made the marks at the Minesota Mine (27–28). There was no material evidence to support any of these speculations.

But questions of antiquity reminded Whittlesey of his hyperskepticism. Calling such concerns "a question of great interest," Whittlesey used the same logic as Squier and Davis; succession to climax vegetation took a very long time. He concluded, "Is it going too far, on the strength of this evidence, to place the *abandonment* of the mines at a distance of 500 to 600 years from our times?" (28).

With no apparent concern for self-contradiction, Whittlesey's closing statements were modest, conservative, and did not go beyond the actual evidence at hand. "There may have been inhabitants covering large territories for long periods who have disappeared without leaving any monumental evidences of their occupation" (28). Ancient people used copper, he concluded, which they extracted in a rude way through shallow excavations. With the simplest mechanics they pounded it free without resorting to metallurgy or cutting. Such tools as chisels and adzes were produced by cold hammering small lumps or masses rather than large ones, for which they had little technical capability to use. They practiced no cultivation and left not much evidence of occupation other than weapons for hunting and defense. "They had to have been numerous, industrious, persevering, and occupied the country a long time" (29). With his two very different scenarios of prehistoric life in the copper region, Whittlesey left the door wide open for continued speculation about the identity of the miners, having not resolved the question at all, despite his considerable contributions.

Winchell the Popularizer

Professor and geologist Newton H. Winchell (fig. 10) published his account of Isle Royale's ancient mines in the *Popular Science Monthly* in 1881, taking on the endurable bogeyman of mining-moundbuilding relationships. His contributions to accumulating knowledge were several. To begin, he examined thin sections of the rocks used for hammerstones and suggested

Fig. 10. Newton H. Winchell. (Photo from Merrill 1924:434, courtesy of Yale University Press.)

that the degree of grooving on a hammer apparently varied with its mineral makeup (Winchell 1881:605). In other words, Winchell framed a testable hypothesis.

The most powerful assessments of the links among moundbuilding, mining, and the contemporary Indian populations were attributable to Winchell. In a carefully structured argument, he showed that there were no discontinuities in the cultures of the moundbuilders, miners, and Indians. The great many references to the use of copper by the native peoples of the continent, and the actual distribution of the copper, yielded this conclusion: "They are far too numerous and circumstantial, and are spread over too wide a stretch of time, to be supposed to be exceptional" (611). He continued in a rather cautious manner. "It is not presumed that this is a complete list of historic references to the use of copper and copper mining by the Indians, but it is amply sufficient to show that it is not necessary to invoke a strange race, prior to the Indians, to account for all the copper that has been found in the mounds, as well as for those found on the surface of the ground throughout the northwest" (614).

Drawing on further lines of evidence, including the continuities in stylistic attributes among native technologies and sculptures, and a worldwide survey of moundbuilding, Winchell concluded that the agency of the native people completely accounted for moundbuilding as well. "From the foregoing it appears that every known trait of the mound-builder was possessed also by the Indian at the time of the discovery of America. It hence becomes unnecessary to appeal to any other agency than the Indian. It is poor philosophy and poor science that resorts to hypothetical causes when those already known are sufficient to produce the known effects. The Indian is a known adequate cause" (619–20). It was the most carefully developed argument of the time regarding the true identity of the miners.

Packard Studies Copper in Worldwide Context

The quality that set R. L. Packard's work (1893) apart was his interest in evolutionary questions of cultural development, a rather trendy emphasis for students of culture history at that time. Steeped in the intellectual tradition of asking big questions about regularities in the course of human history, Packard examined, in the pages of the Smithsonian *Annual Report,* whether copper metallurgy always preceded other metallurgies, and whether it was universally the first of the metals to be utilized worldwide. Was that priority somehow a springboard to further civilization?

He began by analyzing the geographic distribution of copper implements and discovered that because of its proximity to the source, the distribution of glacial drift, and the zeal of local collectors, Wisconsin was the apparent center of copper working in North America. He accepted the notion that mining occurred sporadically over many years and demonstrated that the people were permanent residents of the region, based on evidence of burials discovered during the dredging of the Portage Lake Ship Canal (184). Asserting that prehistoric mining goals were not those of a market economy, he compared the mines and their productivity to other prehistoric mining systems and found them consistent and compatible in scale and technology. "All give evidence that the natives of the country were close observers and possessed a considerable degree of skill in detecting and obtaining the various minerals which pleased their taste or were of use in their simple lives" (189).

He then took on the question, left lingering by Whittlesey, of whether the people resident in the country at the time of the European invasion knew about mining, how to mine, or where to find the copper lodes. He made the very important point that, by 1824, the local native people had been trading with the Europeans for many generations and had apparently replaced their native metals with those they considered more useful than copper (such as iron) (179). Examining a vast body of literature, Packard made the following conclusion. The same Indians who said they knew nothing of mining and were most reluctant to disclose secret information about minerals could, if persuaded, take the French right to the source spots, they well understood the distribution of the material, and they were thought to resort to the mines to glean what they could (190). "Like this old chief, Father Dablon's Indians showed full traditional knowledge when they told him of the mineral localities where, several generations before, copper had been extensively dug" (192). Granted there were no eyewitnesses to mining. Instead there was an unbroken tradition of knowledge about copper in the Upper Lakes, which began with the earliest recorded times and continued until the Treaty of LaPointe (1842–43) transferred the mineral lands to American possession. Packard identified the reluctance to reveal secret knowledge to whites as a long-lived tradition that had only recently (1893) met its demise (191). The word "mining" was in fact a misnomer, because no one ever blasted through bedrock nor dug rock tunnels (192). Packard concluded that confusion about terminology added to the contradictions about mining, which the Europeans reported as savage ignorance.

Packard's initial question, that of the universal priority of copper in the world's metallurgical development, was answered by his assertion that true metallurgy, based on smelting technologies, never developed in central North America. This situation derived from an adequate supply of native (elemental) copper, no ready source of tin for alloying, and the disruptions of cultural change brought about by the European presence.

"Desiring to Examine for Myself the Existing Traces of a Great Native Industry, I Resolved to Undertake a Trip to Isle Royale"

William H. Holmes's straightforward goals were motivated by the need to collect antiquities for display at Chicago's 1892–93 Columbian Exposition, which celebrated the four hundredth anniversary of the arrival of the Spanish in America. He was specifically involved in the development of a comparative exhibit about the gathering of minerals by native peoples and came to Isle Royale and the Lake Superior district to develop maps, photos, and artifacts for his display (fig. 11). All three goals were accomplished; later he published his story in the *American Anthropologist's* 1901 volume. Holmes's account of his method of data gathering on ancient pits was straightforward; he simply excavated one to see what it contained and how it was created. Although he was not the first to do so, he was the best prepared to understand the methodologies of mining in a comparative archaeological framework.

> Wishing to examine the ancient pittings more in detail, I searched the walls of the modern mines for a favorable exposure in which to begin excavation, and finally selected a spot where the complete section of

Fig. 11. William H. Holmes at Minong Ridge, Isle Royale, Michigan, ca. 1891. (Photo courtesy of National Archives.)

an ancient mine, some ten feet in depth and probably twenty feet in diameter, was exposed in a steep slope. The ancient pit was filled nearly to the top with well-compacted material, mainly crushed traprock and earth, the debris of excavation from this and neighboring pits. The most notable feature of these excavations was the ever-recurring sledge-hammers. (Holmes 1901:691)

Holmes's dedication to the comparative method extended to all of the materials he collected. The artifacts Holmes recovered on the island and elsewhere in the mining district were completely familiar given the material cultures of the native peoples of the Lake Superior basin. "The articles mentioned earlier are just such as would be used and lost or abandoned by aboriginal workmen, and serve to connect the known tribes of the lake region with the working of these mines" (695). His commentary on the ubiquitous hammerstone is of interest, because he carefully observed the hammers themselves and discovered some of their very subtle attributes, something no one had discovered.

I conceive it to be quite possible to successfully withe-haft without grooving, and the fact that in many cases there is a polished band around

> the implement at the point where the withe would encircle it, seems to warrant the conclusion that hafting was common. It would look like a waste of energy to undertake the tedious task of pecking a groove in these bowlders [*sic*] when the first blow struck in the quarry work might shatter the implement, making it entirely useless. We observe that the bowlders chosen were generally ovoid in shape and often with the sides approximately parallel, so that withe-hafting would be easy. (694)

Comparative studies of mining pits on the Ontonagon portion of the mining district led Holmes to conclude that there were no important or even perceptible differences in the mining technologies across the region, and thus suggested that different areas were not mined by different peoples, nor at different times, at least based on material evidence at hand.

> At Rockland, near Ontonagon, where a vast amount of modern mining has been done, I found the aboriginal evidences quite plentiful, and apparently identical in character with those of the more northern district. The sledges differ in being more frequently grooved, but this need not be attributed to the usages of a different people, but to the fact that the shapes of the available stones were not well adapted to hafting and had to be more or less completely remodeled to make them available; besides, the material is less brittle, making them better worth the trouble of groove-hafting. (696)

Holmes's work left today's archaeologists a legacy of data collection through field observation based on careful excavation and record keeping and comparative studies of feature and artifact variability. In many ways his report was thoroughly modern in content and was a fine example of basic regional research.

Big Capitalists Join Search for Ancient Miners

A generation later, in July 1928, new participants joined their wallets and hobbies together with those of archaeologists to research the ancients on Isle Royale. The new players were bigwigs from Chicago and Milwaukee; the bank rollers were none other than the president of Zenith Radio Corporation, Eugene F. McDonald, Jr. (a.k.a The Commander) and the vice-president of Colgate-Palmolive-Peet, Burt A. Massee. They sailed around Lake Michigan and Lake Superior from Chicago to Isle Royale in the palatial yachts *Naroca* and *Margo*. Stopping over at Milwaukee, they gathered on board George A. West, a long-standing avocational archaeologist and president of the board of trustees of the Milwaukee Public Museum (fig. 12). The museum endorsed the privately funded expedition. One of the participants was the champion flycaster of the state of Illinois, a longtime friend of Massee's. This was

Fig. 12. George A. West. (Photo courtesy of the Milwaukee Public Museum.)

truly archaeology from another era: a reprise of the dilettante days of the mid-nineteenth century, when philanthropy American style took on its classic shape.

The goals of this adventurous expedition were notable; the group was specifically looking for evidence of mining by prehistoric Aztecs, Norse, or ancestral Indians on the island (West 1929:20). Their general questions were also a reprise of the past: where and when were the people who mined visiting the lake basin? What evidence of occupancy and ethnicity was there? Through site survey the miners' culture was to be discovered, and the expedition would lay groundwork for the future. The Milwaukee Public Museum researched its own copper mining exhibit at Isle Royale in 1924 and held an interest in the archaeology of the region for years, as the primary institution of scholarship in archaeology north of Chicago.

The Norse goals failed, however, to get in the way of objective assessment of the evidence. West's contemporary, the reporter William Ferguson (1923, 1924) implicitly helped to set the McDonald-Massee goals; Ferguson believed that substantial concentrations of pit houses lay buried and disheveled on the island. West and his group were demonstrably skeptical of Ferguson's evidence.[8]

> The long trenches that Ferguson reports were considered by us as natural depressions, although there were several short ones, probably dug for

the purpose of uncovering the native rock and others made in the loose debris of glacial deposits in pursuit of float copper. Hammer stones and mauls which were doubtless taken from the remains of a nearby ancient beach line were numerous. . . . That the pits and drifts at this place were made solely in search of float copper and not for dwelling places seems certain. (West 1929:26–28)

Likewise West judged Ferguson's conclusions about the role of fire in mining in a strictly parsimonious fashion, preferring to demonstrate that when natural causes adequately accounted for material phenomena, they must be forwarded as the most likely causes for them. Speaking of char in shallow pits at Ferguson's "Old Town" surrounded by heaps of spoil rock, West reasoned carefully: "Successive forest fires in centuries past may have distributed charcoal throughout the various strata of soil accumulations found in the pits. Such fires are sometimes very hot and might crumble rock with which they come in contact, even in the bottom of a pit. Consequently, the finding of charcoal or stones, showing the effects of fire in mines of this class means but little" (27).

Thus dismissing Ferguson's claims about buried pit house cities, the researchers proceeded onto the Minong Ridge at McCargoe's Cove, where there was very dense evidence of prehistoric mining. As an amateur archaeologist of some experience, West demonstrated his methodological skills by supporting his generalities with material evidence. His firsthand experience in the copper district began as early as 1875, at least in the Keweenaw (43), at which time he first observed severely disturbed prehistoric pits.[9] Speaking of the Milwaukee Public Museum's excavations of 1924 at pits at McCargoe's Cove, he related that the context or association of artifacts, one with another, was for him the only source of valid inference about the activities of prehistoric people. "In those excavated, much charcoal and many broken mauls were encountered, fifty-six being taken from a single pit. These and other observations clearly indicated that the primitive method of mining by the use of fire, water, and pounding was employed" (3).

Later, at Minong Ridge, West confronted a profoundly sticky myth about the mining evidence, which had still not been vanquished by careful research. It had to do with the work habits of the ancient miners, which to twentieth century capitalists apparently seemed rather perplexing. "Mr. Massee stopped short and said, 'When did this race of craftsmen lay down their tools and depart, who were they, from whence did they come and where did they go?' My answer was, 'We know not. Here is the open door, science is welcome to enter with its traditional tools, the lowly pick and shovel, to explore at will and this work seems to be for us'" (32). Notice that it was the capitalist Massee, rather than the archaeologist West, who was perplexed by the work habits and motives of the miners. No doubt the atmosphere of pre-Depression management-labor relations may have helped to condition Massee's perceptions of ancient work habits.

Life in the shadow of corporate fat cats was entertaining in that day; McDonald was an avid amateur explorer who had been with Macmillan in the Arctic and was fond of recorded images and sounds, which were regularly available on the yachts. After long days of following the evidence of the ancients around Isle Royale, the expedition members resorted to the boats for food, rest, and entertainment in comfort, not to mention the opportunity to take part in the volatility of the Roaring Twenties stock market.

> Although the work during the day was unusually strenuous, our evenings were made most profitable and entertaining, both of the yachts being provided with victrolas and moving picture projectors; the *Naroca* showing, among other things, very interesting reels of Arctic scenes; the *Margo*, the sinking of ships during the war by German submarines. The *Naroca* had aboard a short wave broadcasting outfit and receiving set which kept us in touch with the outside world, enabled us to receive market quotations and to send reports to the press and other messages. (34–35)

One evening the expedition sent a radiogram from the *Naroca* advising President Coolidge that Isle Royale should be made a national park or monument, proclaiming that "it should be preserved in the interests of archeological and geological science" to help preserve and enshrine its "profound archeological, geological and historical interest, its unique scenic beauty." In response, "a courteous reply was received by Commander McDonald, from the Secretary of the Interior, Mr. Roy O. West" (40–41).[10]

In a fashion popular in 1920s efficiency studies, George A. West assembled what was known of the social and subsistence activities of the ancients into a time-and-motion study of such parameters as the labor required to mine, the probable division of labor, likely task specialties, and so forth. He based this work in bridging arguments from ancillary bodies of data, such as the environmental characteristics of Isle Royale. He reasoned that the miners were knowledgeable about the physical details of their world and that their mining activities had a local impact on the appearance of the environment. Their occupancy of Isle Royale, at least, may have been limited by what West measured was a scarcity of game other than fish. Miners followed telltale streaks of malachite, a mineral associated with copper veins and fissures, and the activities of mining the copper consumed fuel and contributed to denuding the local vegetation. Some miners may have organized gangs to procure and transport rocks for hammers from the beaches, which were littered with raw materials identical to those of hammers in situ: diorite, granite, greenstone, porphyry. Other miners may have served as water carriers if mining with fire was widely practiced (56). In all, West's attempts at inference were shot through with material references, with arguments based on association and archaeological context, and with supporting evidence drawn from local ecological and environmental conditions.

West grappled with another of the most formidable copper miner myths. This one had to do with calculations of how much copper, in weight, was mined by the prehistoric people. West estimated "that the weight of all copper implements and ornaments, made of Lake Superior copper, thus far recovered, does not exceed two tons, which, compared to the great amount of work done at the mines, is insignificant" (60). His conclusion? Most copper artifacts have yet to be recovered, and the total number must be vast, he reasoned.

The final question, and the one that prompted the expedition in the beginning—the ethnic identity of the miners—was finally confronted. Unlike earlier writers and researchers, West had the benefit of comparative knowledge not only of archaeology but also of the ethnology of the region. He was familiar with the territories, material culture, and economic organization of historic-period ethnic groups such as the Chippewa and Sioux. At the outset, he clarified his position regarding the potential agency of the Norse: "Students of archeology are almost unanimous in the belief that they were Indians and none but Indians. That they belong to one of the two great families, the Siouan or the Algonquian, seems to be quite certain" (64).

West's model of prehistoric resource access may have sprung from his own immersion in a modern market economy based on differential access to commodities; he favored what one could call a "single-tribe strong-arm control model" of access. That is, he reasoned that one small group managed the copper resource and accessibility to it and restricted all others from it. This model has been widely debated in the anthropological literature of the region in more recent times (Clark 1991); many researchers concluded that it was at odds with what was known of prehistoric sociopolitical organization in the region. West demoted the Chippewa from this role, based on the fact that European and American historical accounts identified them as newcomers to the lake region themselves. "More than that, the Chippewa knew nothing of mining and had no tradition in regard to it. They, like the Indians of all tribes, admired the shining nuggets and treasured them as divinities" (West 1929:65). Like other researchers of his time, West apparently assumed that divine status somehow ruled out a practical utility for copper and that reluctance to divulge mining lore was evidence of ignorance rather than defense of traditional belief.

In fact both the Sioux and the Chippewa claimed the territory of the western basin as their own: the Chippewa on the basis of conquest and the Sioux on the basis of long history. To West, however, one fact stood out for special notice. The treaty "territory claimed by the Menominees and used by them as their hunting grounds, happens to be the district of Wisconsin where the greatest number of copper artifacts and workshops have been found" (66). West therefore favored the preeminence of the Menominee and supposed a link with their Algonquian-speaking ancestors as the main miners and fabricators. He demurred such a pronouncement in the face of vast reaches of time and the likelihood that many tribal cultures had come and gone (67). In the end, he, like so many others, pronounced the question still open.

Drier and Du Temple Assemble Disparate Sources and Theories

In 1961, Dr. Roy W. Drier, metallurgy professor at the Michigan College of Mining and Technology[11] and his colleague Octave Du Temple published a collection of what they found to be seminal articles about the history of prehistoric copper-mining research (fig. 13). These articles, originally published in scattered journals and more obscure formats, covered three centuries of pondering about the ancient miners (Drier and Du Temple 1961). The purpose of Drier and Du Temple's efforts was to make these bits of information accessible to those who might be interested in prehistoric mining. Resulting consumer response was favorable; the book proved immensely popular and salable and by 1965 had entered its second printing.

The principal questions that enthralled Drier and Du Temple were two; the first had to do with the absolute antiquity of the prehistoric copper industries of the Lake Superior basin, and the second had to do with copper's dissemination to prehistoric users. These problems absorbed Drier from the 1930s onward. As a metallurgist, Drier wanted to devise ways of identifying the geological sources of copper in sundry metal artifacts found elsewhere around the world, as a means of tracing the dissemination of raw copper across continents. The discussion confronted the notion that there was a system of worldwide trade in which American copper artifacts were popular commodities. According to these researchers, "there is no evidence that links the Lake Superior region with Egyptian copper, but as of this date the Egyptian civilization is definitely one that must be considered. It is estimated that 500 million pounds to perhaps more than one billion pounds of copper were mined prehistorically in the Lake Superior area. Where this copper went is still a mystery" (15). The duo suggested that this trade likely went beyond Egypt, and existed far earlier than previously imagined. "It

Fig. 13. Roy W. Drier. (Photo courtesy of the MTU Archives and Copper Country Historical Collections, Michigan Technological University.)

is certainly possible that a copper trade existed several thousand years ago between America, Europe, Asia, and South America. Improved analytical techniques, dating techniques, and knowledge of the geochemistry of copper may help us to trace the routes and determine to what extent copper moved in world trade in prehistoric times" (15).

Drier was somewhat restrained in his earliest speculations about the ethnic identity of the miners; these writings dated to about 1935 (31). In addition, he routinely discounted the veracity of other people's speculations. For example, he rounded down by an order of magnitude early estimates about the imaginary size of the prehistoric labor force involved in mining. But he was inclined to support the notion that multiple cultural groups did the mining. Drier's goal was to document an American Copper Age as substantially as the Europeans had their Bronze Age, and to create a bridge in the inevitable transition from the Stone Age to an Age of Metals. In this pattern Drier was very much influenced by nineteenth- and early twentieth-century views of inevitable world progress, in which it was necessary not only to have an Age of Metals but to demonstrate its unquestioned improvement upon Stone Age life. To Drier, the transition to metals was profound and irreversible and happened in the same order everywhere. This inevitable progressive transformation yielded traits such as increasing complexity of maritime technology, tools, and material culture. For instance, grooved hammers, in this view, were of necessity a technological improvement over ungrooved ones. Thus, when describing mauls or hammers in use by the ancient miners on Isle Royale, Drier put it this way: "None, or very few of the stone mauls found on Isle Royale have grooves. Certainly, the grooved hammer-stones are evidence of a later stage of civilization or at least a reversal on the average slope of the civilization curve" (23). Embedded in Drier's posing of important questions were two unwarranted assumptions: that there were at least two groups of miners, the first group mining Isle Royale, followed by more civilized people who exploited the mainland using highly evolved grooved hammers (30). The second was that the miners, in leaving no trace of other trappings expected of evolved civilization, presented evidence for their own abrupt and unplanned departure: "Who these miners were, where they came from and why they left their operations as though expecting to return the next day—but didn't—are the questions which are still bothering archaeologists" (30). Drier took some of this mythology to heart: "Some of the pits were only half worked and some of the old mining operations were found with the tools left in them, just as though the miners had quit for the day, and planned to return" (23).

Drier's later accounts, published in 1953, although disclaiming some speculation, still attributed an uncanny ability to the "old timers" regarding their successes at prospecting. "What native talent or ability enabled them to locate copper-bearing lodes even though the formations were buried?" (23). Regarding the total take of prehistoric copper, he cited "early investigators" who suggested recovery of "from one million to one and a half million pounds of copper" (21). Drier then established, to his own satisfaction at least, that there were no local sources of mining hammers on Isle Royale, an erroneous statement (Bastian 1963) that set the stage for exotic miners to have control over the mining. Drier's accounts were replete with "clouds of mystery" and "veils of magic" (Drier and Du Temple 1961:23), an orientation which, rather than that of Drier's occasionally practiced skepticism, took hold in the minds of his readers and refused to let go its sticky grip.

In 1953 Drier arranged funding from Joseph C. Gannon of Marquette, Michigan, from a scholarly foundation and additional support from Michigan College of Mining and Technology to undertake archaeological excavations of a prehistoric mining pit at Isle Royale National Park. Drier was trained as a metallurgist rather than as a field archaeologist, a distinction he presumed did not really stand in his way. "Tho I am a metallurgist, I have read much archeology, and a report that I submitted to the Director of the Milwaukee Museum, according to him, was more worth of a doctorate in archeology than anything he'd seen for a long time" (National Park Service 1953). In the same communication, Drier presented his assessment of the archaeological potential of the Isle Royale pits.

Before the Island was a park—and before radioactive carbon was known—I dug into some of the old pits looking for hammer stones and pieces of copper. I did not find any hammer stones in the pits, nor any copper other than small pieces from the vein. There was nothing in the pits of archeological significance. These old pits have filled in with rock broken loose form the bed rock by the action of water freezing in cracks and by the splitting action of roots, and also filled in with humus and roots. The roots and the fine pieces of rock make digging a miserable job. (National Park Service 1953:Drier to McLaughlin, March 11)

The National Park Service, reluctant to issue an antiquities permit to an untrained archaeologist, insisted that Paul L. Beaubien, the Park Service archeologist assigned to the region, accompany Drier and his Michigan College of Mining and Technology field party to the island, to which Drier agreed: "I would be delighted to have your Mr. Beaubien accompany us, tho I can see no archeological advantage" (National Park Service 1953:Drier to McLaughlin, March 11).

James Bennett Griffin to the Rescue

Exposure to professional archaeological analyses, particularly those conducted by the University of Michigan's James B. Griffin, altered Drier's later views (fig. 14). Drier encountered Griffin in 1953 on the first of several Michigan College of Mining and Technology Isle Royale expeditions, which Drier organized to gain samples for dating by the (then) brand-new radiocarbon method. The resultant radiocarbon dates, some in excess of 3000 BP, confirmed that the mining pits were truly of prehistoric age. Drier's reports of these trips were in part related in Griffin's *Lake Superior Copper and the Indians* (Griffin 1961). With the appearance of James B. Griffin on the scene and the supporting evidence of radiocarbon estimates of age, speculation was no longer the center of the process.

Griffin's participation in the archaeology of the copper-producing regions of Michigan began in 1953, when he joined Drier and staff on Isle Royale. Beaubien's field report of 1953 recorded this event: "Dr. Griffin, Director of the Museum of Anthropology, University of Michigan, requested permission to accompany the party a few days before the group left Houghton, and was cordially welcomed by Dr. Drier. Dr. Griffin will have the samples of charcoal dated without cost, and he can facilitate other of Drier's research projects which require the analysis of artifacts from widely separated sources" (Beaubien 1953:[1]).

No stranger to Upper Michigan archaeology, Griffin's Isle Royale research interests complemented the lengthy research achievements of the

Fig. 14. James B. Griffin. (Photo courtesy of the University of Michigan Museum of Anthropology.)

University of Michigan Museum of Anthropology staff in the north. Carl Guthe and archaeologist Fred Dustin worked on Isle Royale during 1930–50; Albert Spaulding followed in 1953. Fieldwork in the summers of 1956, 1957, and 1959 by Griffin in collaboration with his colleague George I. Quimby of Chicago's Field Museum of Natural History and student Mark Papworth of the University of Michigan Museum of Anthropology tracked copper production and utilization into the eastern Lake Superior drainage (Papworth 1967). The results of Griffin and Drier's expeditions were eventually published in *Lake Superior Copper and the Indians* (Griffin 1961). Griffin interpreted both the artifactual evidence and the time-space systematics of several archaeological assemblages, then linked recovered artifacts with the complex prehistory of the region. Drier completed the write-up of the 1953–54 Michigan College of Mining and Technology excavations and added an overview of comparative metallurgical work.

These excavations were quite modest in extent. The goal was the excavation of Pit 25 on Minong Ridge to determine the age and typical contents of a prehistoric mining venture. This pit was positioned by the miners to intersect a fissure running parallel to the ridge, in which were relatively rich deposits of copper. The excavation was conducted in one-foot levels; by the end of the 1954 season it was fourteen feet deep (6). The pit deposits appeared completely undisturbed by more recent events, and radiocarbon samples dating to 3000 +/– 350 BP, 3800 +/– 500 BP, and 3310 +/– 200 BP were recovered (M-320; M-371e; W-291; Griffin 1961:5–6). Drier documented the location of what he referred to as "a prehistoric stamp mill" several feet removed from the pit and commented that the surrounding debris of small hammers, bits of trap rock and copper, and worked copper suggested that the location "was an area where the prehistoric miners crushed the rock which they had removed from pits" (7). Drier concluded that the estimates of pit numbers repeated since the late nineteenth century were within reason: "the 1870 estimate of three thousand pits and one thousand tons of hammers does not seem to be out of line" (7). In this work, speculation as explanation was somewhat eschewed by the influence of Griffin, not one prone to call upon Egyptian or other alien prime movers to conduct the affairs of North American prehistory.

Griffin's most entertaining contribution to the volume was his challenging of the dual bogeymen of myth and speculation, in which he concisely debunked rampant ideologies of hyperdiffusion and nineteenth-century racism (130–33). With steely logic and merciless fact, he toppled ghosts of lost tribes, lost alchemies, and lost artifacts, and one would hope that by such erudition he had delivered these shades their death blows. But the University of Michigan was not quite ready to withdraw from Isle Royale.

A Comprehensive Look at Mining Technology on Isle Royale

Tyler Bastian's three years of work on Isle Royale through the University of Michigan Museum of Anthropology represented the most comprehensive body of work on prehistoric copper mines and mining from a field perspective in the Upper Great Lakes (Bastian 1963; Clark 1995:5). As a result, the level of archaeological work on Isle Royale still exceeds that conducted on the mainland. Ontonagon County still awaits comprehensive archaeological survey to recover data on prehistoric copper mining.[12] The ancient copper workings on the northern Keweenaw Peninsula have yet to be researched systematically. So Bastian's Isle Royale work constituted the initial field-based attempt to understand the technological context of ancient copper exploitation. Clark's later work studied the social context of copper use in the Lake Superior drainage (Clark 1991, 1995).

Bastian worked through University of Michigan Museum of Anthropology for three field seasons on Isle Royale, 1960–62. His emphasis was on surveying and testing reported sites on the coast and in the interior. The outcome was a summary of what was known of Isle Royale's prehistoric mining, its age, technology, and links with the rest of regional prehistory

(Bastian 1963). He attempted to evaluate prior work fairly in the light of clear logic and good archaeological scholarship.

Among Bastian's many accomplishments on Isle Royale were his descriptions of the sorts of copper deposits systematically exploited by prehistoric people and the special qualities that usable copper deposits shared. Prehistoric miners selected certain modes of deposition for working, specifically the amygdaloid deposits (18). At places where fissures intersected each other, or cross lodes, pit-like excavations often appeared in disconnected linear patterns. Bastian's standards for what features actually constituted prehistoric mines were conservative and stringent, but resulted in systematic description of the evidence. "Only reports of mine sites confirmed by the presence of unquestioned hammerstones or by excavation of one of the suspected pits at the site can be accepted as reliable" (21). He established these criteria as his working definitions of real mining evidence and, as a result, developed a conservative estimate of the numbers of mining pits, ($n = 1089$) based on rule satisfaction and stringent definitions of terms rather than freehand estimation. He showed that there was a great deal of consistency in the appearances of the excavated mines. Sometimes they included organic materials, as well as artifact forms familiar from other prehistoric Native American sites. Wood materials were interpreted by the most conservative means; that is, unless natural causes such as forest fires could be ruled out, the agency of humans was not regarded as a powerful explanation for the data. Bastian found no evidence of water drains in the pits the crew excavated, nor was there a great deal of charcoal and ash in them.

Bastian's interest in experimental creation of mining evidence was also enlightening, for he tried to replicate the processes of mining, then examined critically the products of his work to evaluate some of the conclusions others had developed regarding the sequence of mining activities and the relative physical effort involved. He was particularly curious about the efficacy of mining with hammerstones, the role of fire-setting in the mining effort, and the amount of time required to gather copper trapped in bedrock.

Bastian's summary about the scale of Isle Royale mining modeled the gradual accumulation of mining debris over thousands of years of mining activity, a time period over which the requirements of the native cultures of the region adequately accounted for the body of prehistoric mining evidence. Mining over many years' time by a few people was a process compatible with the lifeways of the regions' native peoples and provided more than adequate numbers for accomplishing the mining on the island, through "the cumulative result of small contributions" (58).

Examination of occupation sites on the island and the presence there of worked copper fragments indicated a long-lived use of Isle Royale and the continuity of copper acquisition and use up to and throughout the time of the Middle Woodland cultural tradition and later times. "The broad pattern of the use of copper in North America is a generally continuous one from about 4000 B.C. to the historic period, with the main centers of copper consumption moving southward. It seems reasonable to suppose that the mines of the Lake Superior region supplied most of this copper" (58).

His final remarks suggested that the Lake Superior basin mining cultures were not in any important way different from those encountered by the first Europeans in the area, and that what uniqueness might be attributed to the area was in large part due to its unusual deposits of copper rather than to any out-of-scale highly exotic procurement strategy. In the end, Bastian provided systematic description, clear definition of terms, adequate experimentation, and field recovery of data to interpret archaeological evidence.

What Has Been Learned?

Material evidence of the past and speculation about it are not the same. Despite one hundred and fifty years of systematic research no one has found one scrap of evidence to show that anyone other than regional native peoples ever mined copper in the Upper Great Lakes. Virtually every serious writer and investigator has for more than a century of consistent reporting given no credence whatsoever to the mythology of trans-Atlantic trade in copper. Nor have they given any support to stories of long-distance trade or migration of peoples to and from Central America to Lake Superior. By processes of gradually accumulating knowledge, innovative developments in scientific methods of analyses, and persistent refinement of research questions one hundred and fifty years deep, there is widely shared irrefutable evidence that native American peoples mined copper over long reaches of time, ending their activities somewhere around the time of European contact.

Questions of the relative antiquity of the mining activities, the ethnic identities of the miners, their links within and outside of the mining region, and their technologies and trading ties have been repeatedly posed and steadily answered since 1845. Throughout these decades the scientists in search of the answers developed and refined most of the basic logical, theoretical, and methodological approaches on which contemporary archaeologists routinely rely. Many of the basic researches about copper mining contain surprisingly modern approaches to archaeological inquiry; in fact the story of this research recapitulates the story of the development of modern American archaeology. These researchers developed and applied such concepts as argument by analogy, the necessity of understanding archaeological phenomena within a research universe, the requirements of the

Law of Superposition, the priority of in situ evidence, the importance of experimental replication, and the need for explanatory parsimony. Scattered throughout the story one finds reference to the basic analytical techniques that have become standard procedures for archaeologists: thin-section analyses, radiocarbon dating, reliance on dendrochronology, and understanding formation processes. These approaches are complex and time-consuming, but are fundamental to modern archaeology and are the only approaches that can consistently produce reliable comparative data for productive research.

Though archaeologists now know how to go about gathering the evidence they require, their ability to understand it adequately is still incomplete. Some of the critical data have yet to be recovered. For example, copper workshop sites should eventually inform archaeologists about how and where the copper materials were processed into artifacts. Though these kinds of sites have been sought, few have actually been found and fewer still excavated (S. Martin 1993). Both the Keweenaw Peninsula and the Ontonagon River drainage require additional systematic treatment, data recovery, and interpretation on a regional level before more developed questions about copper mining can be addressed and reasonably answered.

Suggestions for Further Reading

Griffin, James. B., ed. *Lake Superior Copper and the Indians: Miscellaneous Studies of Great Lakes Prehistory.* Anthropological Papers 17. Ann Arbor: University of Michigan Museum of Anthropology, 1961. A comprehensive account of material evidence for early mining and prehistoric occupation of the Lake Superior region, with a focus on Isle Royale and the north shore of the lake.

Holmes, William H. "Aboriginal Copper Mines of Isle Royale, Lake Superior." *American Anthropologist* 3 (1901): 684–96. Holmes's brief account of his initial trip to Isle Royale and his investigations there.

Squier, Ephraim G., and Edwin H. Davis. *Ancient Monuments of the Mississippi Valley.* Smithsonian Contributions to Knowledge 1. Washington D.C., 1848. The classic account of the American antiquities of the central part of the continent, with beautiful engraved illustrations.

Whittlesey, Charles C. *Ancient Mining on the Shores of Lake Superior.* Smithsonian Contributions to Knowledge 13, no. 155. Washington, D.C., 1863. Pp. 1–29. Meticulous descriptions and drawings of prehistoric mining evidence, including maps, geological cross-sections, and artifact sketches.

Willey, Gordon R., and Jeremy A. Sabloff. *A History of American Archaeology.* 2d. ed. San Francisco: W. H. Freeman, 1980. A very readable overview of the development of scientific and academic archaeology in North, Middle, and South America.

4

Of Heaps of Rubble and Earth along the Courses of the Veins

Early Evidence of Copper Mining

> The evidence of the early mining consists in the existence of numerous excavations in the solid rock; of heaps of rubble and earth along the courses of the veins; of the remains of copper utensils fashioned in the form of knives and chisels; of stone hammers, some of which are of immense size and weight; of wooden bowls for bailing water from the mines; and numerous levers of wood used in raising the mass copper to the surface.
> —John W. Foster and Josiah D. Whitney, 1850

Introduction

The native peoples of the Lake Superior region learned to locate, extract, and process native copper to ready it for uses as artifacts, ornaments, and tools. To appreciate how this was done, the stream of activities undertaken—from discovery, to acquisition or extraction, to eventual alteration into final form—must be clearly described. This approach, one of accounting for all of the activities, processes, and ancillary materials (including energy sources and local landscape impacts) that are necessary to produce an artifact or object, is used both in understanding prehistoric mining (Ericson 1984; O'Brien 1994:12) as well as in studying industrial-scale mining (Noble and Spudy 1992). The latter refer to this approach as "property analysis"; O'Brien is satisfied with the term "metal production system," but in fact all of these authors are referring to the same intellectual exercise. Regardless of name, the important quality of studies such as these is that they be comprehensive and account for all things, places, and, most important, the *social units* that affect or are affected by a production flow of a single resource. By the close of this chapter the reader will have, figuratively speaking, walked through all of the stages of metal acquisition and will be somewhat familiar with its material elements and physical evidence. These activities include prospecting for bedded copper, collecting loose copper, recovering copper from veins in rock, using energy sources, using associated tools, raising copper masses, removing water, and removing trap rock waste. When these activities, their labor/energy requirements, and their archaeological consequences are understood, it may be possible to make informed statements about the kinds of cultural groups who undertook them.

Discovering Copper Sources

Back in the days when the evidence for prehistoric mining was first discovered, reporters found it puzzling that ancient people could have discovered copper deposits, which were buried and hidden from direct observation, with what seemed to be unnatural levels of accuracy. Ancient abilities to discover the systematic patterning of geological processes were in fact so misunderstood by Euro-Americans that they eventually ascribed to the ancient people some sort of quasi-supernatural talent of prospection. The facts of the matter, however, are quite the opposite of supernatural inspiration or mysterious inscrutability. Nature, whether observed by European or native American, in this century or any other, is composed of patterned occurrences. These patterns are knowable, and the immersion of anyone, in any environs, establishes a particular context in which learning about patterned events inevitably begins. Later, these patterns can be extracted, summarized, explained, and passed on to others through stories, myths, songs, and poems. All peoples become accommodated to their local settings through the processes of pattern recognition and transmission. There is nothing more human than this behavior.

Unfortunately we cannot ask the original miners to describe their skills in pattern recognition, how they passed on accumulated knowledge about mining, or what that knowledge was. But Georgius Agricola, a sixteenth-century European who was the first to write down in detail the kinds of mining activities the people of his culture undertook, was likely the absolute contemporary of some of our ancient American miners. He collected the accumulated verbal knowledge about his culture's mining crafts and put it in a printed book, which was a relatively recent invention of his time. Looking at Agricola's book, *De Re Metallica,* gives us some understanding of how nonliterate people found minerals. Agricola's information developed through experience, which was then transmitted orally, much as we assume that the ancient peoples of our region learned and practiced their skills.[1]

The geological patterns that the miners of Agricola's culture recognized were the kinds of knowledge useful for prospection everywhere. Experienced miners, he said, knew how to find veins from surface indications. Sometimes they were stripped bare of ground cover by chance, through events such as erosion, fire, tree falls, and seismic activity, but these were relatively rare events. Other methods were needed. "Further, we search for veins by observing the hoar-frosts, which whiten all herbiage except that growing over the veins, because the veins emit a warm and dry exhalation which hinders the freezing of the moisture, for which reason such plants appear rather wet when whitened by the frost" (Hoover and Hoover 1950 [II]:37–38). In the same way, Samuel O. Knapp, the individual who first reported the occurrence of ancient mining in the Lake Superior district, discovered his ancient evidence: "[H]e observed a continuous depression in the soil, which he rightly conjectured was caused by the disintegration of a vein. There was a bed of snow on the ground three feet in depth, but it had been so little disturbed by the wind that it conformed to the inequalities of the surface" (Foster and Whitney 1850:159). Knapp then discovered ancient workings in a purposefully enlarged cavern at the terminus of the vein, which included stone hammers and the worked vein itself.

Other clues about the veins were apparent to the skilled observer: discolored vegetation, stunted trees in a path through an otherwise thick stand, unique patterns of vegetation such as fungi and herb appearances in patches or paths—all marked a vein (Hoover and Hoover 1950 [II]:35–38). Agricola also suggested that spring waters were sometimes indicative of metal-bearing veins. Water action of all kinds enhanced the likelihood that copper could be found. First was bedrock erosion. Flowing water or wave action against rock faces revealed entrapped copper, for example at numerous places where water flowed over the steep inclines of bedrock. Water movement also carried copper nodules to sites of drift redeposition downstream. In the Lake Superior region, these banks of drift were systematically exploited by ancient people in a number of localities near the bedded sources, including Isle Royale (West 1929), Portage Lake (Whittlesey 1863) and the tributary branches of the Ontonagon River (Troy Ferone, personal communication, 1996).[2]

Not all of the exploited copper was deeply buried. Glacial scouring removed overburden and revealed embedded veins, resulting in surface exposures. This was certainly true of the Keweenaw (Wood 1907), and Holmes supposed that the same processes revealed veins of copper at Minong, Isle Royale (1901:687). Winchell remarked that at Isle Royale, mines were discovered where accidental outcroppings "of marked lithological character" occurred (1881:601), which were then systematically followed. On the Keweenaw Peninsula, Whittlesey related that "the ancients did not neglect the most trifling indications of metal, but appear to have instituted a thorough investigation as to whether the copper existed in true veins, in metalliferous bands, or in detached nests. There is nothing remarkable in their operations at the "Native" copper and "Northwest" mines, except this closeness of pursuit, through all the veins and branches to their most minute extremities" (1863:6).

Their search was sometimes enhanced by the occurrence of exposed veins of distinctive companion mineralization that were associated with metal-bearing veins. This occurred at many places in the Lake Superior basin. Lakeside traces of such veins, for example La Roche Verte, which extended from the shore into the waters of Lake Superior near Copper Harbor on the Keweenaw Peninsula, signaled the presence of the metal. Similar features occurred on Isle Royale.

Like other patterns, geological processes are knowable and somewhat predictable. They account for the appearance of the land: its terrain, surface composition, and variability. What lies below the surface dictates, to some degree, what appearances the surface will take. Processes of change, such as erosion and weathering, alter surfaces and reveal underlying phenomena. These facts of geology were not lost on the ancient miners. First, the localities of metal beds were not only predictable but somewhat visible on the landscape. Second, the erosion and transportation factors related to the localities of drift copper were clearly obvious and required no uncanny abilities, any more than did understanding the formation of a sand bar at the mouth of a stream.[3] So the "inscrutable copper miner" school of thought was eventually replaced with the idea that people become intimately familiar with their surroundings and become aware of the constituents thereof and the processes that are acting upon them. The locales of the metal were self-evident and the subsequent working of copper a derivative of the technologies that had been used on other rocks since time immemorial.

What Does a Prehistoric Mine Look Like?

Holmes (1901) compared them to the littered grounds of prehistoric flint quarries; Foster and Whitney (1850) and Whittlesey (1863) thought they could be very subtle in appearance. Whatever the case with the individual observers, through a composite story of early historic-period descriptions comes an image of the surface appearance of the prehistoric mines. They were highly variable (fig. 15). The strategy used to open them and collect copper from them varied with the kind of copper deposit they tapped and the circumstances of their immediate topography.

The Keweenaw Peninsula, the Ontonagon River drainage, Isle Royale, and perhaps a few other areas of the Lake Superior basin are the only places in eastern North America in which a person can, today, walk up to a prehistoric copper mine and look at it. The following account derives from firsthand

Fig. 15. Line of aboriginal mining pits at 20IR80, Washington Island, Isle Royale National Park. (Photo by Patrick E. Martin, reprinted with permission of the National Park Service, Isle Royale National Park.)

observations of clusters of prehistoric mines at and near Caledonia Bluff, Ontonagon County, Michigan, around 1991. To get to the pits, one must climb up onto the trap range. This is an exciting climb from the southwest, because the ridge is very steep and handholds are few. Grabbing small trees for support only helps a little; the combination of a backpack, camera, and slippery field boots makes for a precarious climb. Once at the top of a rather narrow and sinuous ridge, one is treated to a panoramic view of the surrounding ground and the courses of nearby rivers.

It is May and the surface visibility is excellent; the local vegetation and bug biomass have just begun to emerge. Everywhere there appear to be irregularities in the surface of the ground, some of which are natural, some artificial, and some a bit of both. The natural terrain is uneven to begin with; there are jagged outcrops, rock faces, and grotto-like intersections of bedrock. Among these are pits and hollows, some roughly two to three meters or more in depth, and a meter to several meters wide. They are spaced irregularly and might occur at intervals of ten to twenty meters, though this is highly variable. They occupy, on and off, the entire distance from the narrow southwest point of the ridge, then northeast a distance of a kilometer or so. These pits are irregular in shape, depending on the natural terrain surrounding them; trees grow out of some and have managed to make their contours less regular.

The ground cover is uneven as well, ranging from bare bedrock to deep deposits of humus and leaves. At places where the ground surface is more visible are found shattered fragments of bedrock. In some places these appear as natural talus-like deposits and in others the material is mounded up and the deposits appear intentional. Scattered throughout the fractured bedrock are other rocks, of different colors and shapes, which appear to be regularly shaped. Some of them are spalled longitudinally; others are broken in half, and still others are whole. They appear to be water-rolled pebbles ca. 2–3 kg in weight, and about 15–20 cm long. Their regular shapes and patterned wear are very distinct and differentiate them immediately from the background trap material (fig. 16). Some are grooved, but most appear plain and unaltered on their surfaces except for scars of wear. They are mining hammers.

Removing surface vegetation, leaf cover, and humus, the pits appear to be chock full of rotted and sometimes wet organic material and fractured rock. Their interiors are also tangles of roots, which have occupied the somewhat loose soil and its spatial voids with enthusiasm. In some pits the fill appears to be dusty and dry, almost ash-like, but most of it, by volume, is detached rock. Moving this material is hard work and goes on to quite a variable depth. These depths may be as little as a meter or as great as three

Fig. 16. Hammerstone scatter, Isle Royale National Park. (Photographer unknown, reprinted with permission of the National Park Service, Isle Royale National Park.)

meters. There appears to be no internal stratification of redeposited material, but the fill appears to be the result both of natural and intentional processes. At the bottoms or along the sides of some of these pits are thin veins of copper and accompanying minerals (fig. 17).

In fact the appearance of this ground today bears great similarity to historical accounts of its nineteenth-century look. Foster and Whitney's famous account of the mining evidence introduces this chapter, and their full text includes an account of the investigations of Samuel O. Knapp, who, after his initial discovery, proceeded to excavate a pit to see what it looked like. He found it to be greater than twenty feet in depth, and still encasing a mass of copper that the early miners attempted, but failed, to excavate. Instead, they hammered every possible face of the mass to free small, usable quantities of copper from it.

Foster and Whitney's account described the appearance of other mined grounds near the Minesota location: "The vein was wrought in the form of an open trench; and where the copper was the most abundant, there the excavations extended the deepest. The trench is generally filled to within a foot of the surface, with the wash from the surrounding surface intermingled

Fig. 17. Exposed copper vein within aboriginal mining pit at 20IR80, Washington Island, Isle Royale National Park. (Photo by Patrick E. Martin, reprinted with permission of the National Park Service, Isle Royale National Park.)

with leaves nearly decayed. The rubbish taken from the mine is piled up in mounds, which can readily be distinguished from the former contour of the ground" (1850:159).

The observations of Foster and Whitney clearly reflect three facts of ancient mining. It occurred opportunistically, the methods used to recover copper varied, and the mines took more than one form. They were highly variable in all dimensions. The 1850 report commented on the geographic extent of the mining: at the Minesota location, in its general vicinity, northeast along the trap range, and on Isle Royale. At Minesota the workings extended sporadically over a distance of greater than two miles; at the nearby Forest Mine location, four veins were worked across a hill slope. Turning northeastward, for "a distance of nearly thirty miles, there is almost a continuous line of ancient pits along the middle range of trap, though they are not exclusively confined to it." Continuing east-northeastward along the Keweenaw Peninsula, they extended "a distance of twelve miles, along the base of the trap" (161). At Isle Royale, the workings "can be traced lengthwise for the distance of a mile" (162).

Alvinus Wood visited many ancient mining pits as early as 1853 or 1854 and then, in 1855, excavated an ancient pit at the Quincy Mine location. He found that "it was shown to be 14 ft. deep, having been filled up by the sliding-in of material composed largely of broken rock, taken out in sinking the ancient shaft, and left near its mouth." The pit, seven feet in diameter, and others like it "could be traced on the lode for a hundred feet or more to the south" (1907:288–89).

Visiting a large number of ancient pits gave Wood an appreciation of their variety and instilled in him an understanding of why some areas had been heavily exploited. Wood's commentaries on the appropriate ground for exploitation and the useful forms of copper for prehistoric uses were of some insight; he believed that for native purposes, "barrel work" was just right. In Wood's industrial jargon, this referred to "copper too coarse to go to the mill, yet not in pieces large enough to be shipped separately" (230). The larger masses, such as the one raised and abandoned at the Minesota location, were too large to divide easily. Tiny granules of copper were too small to utilize, and such deposits, though explored, were not extensively mined. Thus, he suggested that the miners sought out ground where appropriate-sized copper could be found. In addition, their quarries were concentrated in places where overburden was shallow and the copper-bearing deposits were accessible. Pure masses were not neglected, but were hammered to remove any detachable surface pieces; rocky masses were useful and were hammered apart. The Keweenaw's small fissure-veins were heavily worked: "[N]early north of the North West Mine, a number of ancient pits were found on a small fissure-vein, which carried sheet-copper well suited to the uses of the old miners" (233). The interest in fissure veins was apparent at Isle Royale's workings as well, where flat sheets of copper could be useful and "much labor would be saved" (Fox 1911:88).

Whittlesey's extensive tour of the Keweenaw mining evidence offered the opportunity not only to track the geographical distribution of the pits, but to examine the hammered walls of the mining pits and rock faces themselves. The Waterbury mining location presented what he called an artificial cavern, the result of mining activity. Turning to the ancient excavation, Whittlesey "examined the walls of this cavern minutely, hoping to find the marks of some tool of metal. The effects of blows of stone mauls were visible, and such is the hardness of the rock, that if drills or picks had been used upon it, I think the marks would be easily seen, particularly on that part which was protected from the atmosphere by water" (1863:8). He noticed similar maul scars at prehistoric workings at the Minesota location, where a support pillar of trap rock arched over the worked trench, and commented that the overhanging rock there "was cut or bruised quite smooth" and that "the marks of fire on the rocks of the walls are still evident" (18).

At Isle Royale, Winchell observed that "these mines are rude, irregularly disposed, shallow pits in the general surface, which, on being cleared of rubbish, are found rarely to exceed the depth of ten feet, but in some instances reach the depth of twenty. They seem to have been located by the accidental outcropping of native copper, over large areas the rock being entirely bare. In other cases, the mining seems to have been systematically prosecuted along the strike of a known copper-bearing belt of rock" (1881:601).

Holmes concurred that there was little tunneling to be done in this style of surface mining, but the approach used to mine and the shapes of the mines themselves clearly varied with the kind of copper deposit at hand. Later, Bastian codified two kinds of mines found at Isle Royale: fissure mines and lode mines. "The prehistoric copper mines on Isle Royale occur in two general forms, depending on the kind of native copper deposit in which they were dug. The first, fissure mines, are relatively deep conico-ovoid shafts occurring on fissure deposits; and the second, lode mines, are relatively shallow, irregular, quarry-like excavations occurring on lode deposits" (1963:20).

On Isle Royale, the fissure mines were few in number but widely distributed, as were the fissure deposits. They held tabular deposits of copper in thin veins; the deepest of the known fissure mines descended six or more meters. Bastian characterized the appearance of the excavated fissure mines, stating that they "show distinct copper bearing fissures, a clay or sand deposit on the bottom, a compact organic fill containing rock fragments and hammerstones, smooth sides near the bottom, and the absence of a distinct pile of tailings near the mine" (23). Some of the fissure mines contained quantities of charcoal, but Bastian attributed this to natural deposition.

Most of the mines at Isle Royale were the shallower lode variety, and they were found exclusively at Minong Ridge. The pits were variable in size and depth and tended to occur in groups across irregular intervals. The largest known pit mine measured about five by ten meters, and depths varied to a maximum of perhaps four meters. Copper in the target lodes was generally dispersed, in various configurations, in the enclosing rock. Those lode mines excavated were filled with tailings from other quarrying operations and rarely contained much charcoal or ash.

Fissures were the proper size for one person, who could remove tailings with a basket. Lodes were more easily worked than fissures and as a consequence had fewer hammers (51–52). The tailings could simply be tossed out of the working area. Bedrock removal occurred by crushing rather than fracturing, leaving very small tailings; the fines never constituted most of the pit fill, which was generally of angular pieces of trap rock (24). Bastian found no evidence of water drains in the pits, nor was there a great deal of charcoal and ash.

The Processes of Quarrying and Mining

As W. H. Holmes astutely observed in 1901, "the terms 'mining' and 'quarrying,' as applied to aboriginal work, are practically synonymous" (1901:684). By this he meant that tunneling or deep sinuous excavations in the rock did not characterize the prehistoric mining. Rather, open rock faces and shallow surface deposits were exploited and occasionally followed. It appeared that there were at least three different situations for acquiring copper, depending upon the particular kind of deposition at hand: lode, fissure, or glacial drift. Drift deposition basically required finding, sorting, and carrying away copper; Whittlesey (1863) and West (1929) described uneven ground at such deposits, which was likely the result of intentional digging. Whittlesey likened these excavations to gravel pits; they were rich in small copper nodules as well as in the kinds of useful hammers that rock mining required.

The mining process in lode and fissure deposition could be characterized by three stages of activity, and each would likely have been accomplished with particular kinds and numbers of tools. Individual pits might have required ad hoc practical treatment so that each pit had a slightly different associated array of artifacts. Some of the information about mining processes came from historical accounts of ancient mining, and some came from archaeological excavation. Taken as a whole they leave us with a composite picture of what the mining process was probably like.

After locating a likely area and removing loose overburden, mining commenced, a process of finding and following a useful vein or deposit. Not all explorations were successful, for some masses were too large to remove and were abandoned. Hammers in a range of sizes, hafting styles, and weights were the primary tools with which the mining was accomplished. The purpose of *initial hammering* was to create or enlarge fractures in the bedrock. In fissure mines this activity required much mechanical energy. Hammers to accomplish it were probably heavy (> than 5 kg), large (girth >35 cm), and sometimes irregularly shaped. Typically, they exhibited surface rilling or light grooving in multiple locations. Bastian remarked that in his experience it was rather difficult to distinguish such hammers from nonutilized ordinary large rocks. They were likely hafted with flexible thongs, to be swung into the resistant rock. Wedges of stone, wood, or copper may have been driven into the intersection of the vein and surrounding rock

Fig. 18. Experimental mining on Isle Royale, University of Michigan Museum of Anthropology, 1962. (Photo by William Deephouse.)

to loosen the copper. This initial fracturing probably was more demanding of human energy in the resistant rock of the fissure mines than in the looser configuration of the amygdaloid deposits at lode mines. Bastian remarked that there were generally greater quantities of hammers to be found at fissure mines, reflecting the tougher rock and need for more energy to mine it (1963:51).

Being interested in the process of hand mining and in the time and labor required to mine with prehistoric tools, Bastian's field crew arranged an experiment. In the summer of 1962, Bastian, his colleague David H. Thompson, and the crew chose a clean and undisturbed rock face and began the hand mining experiment (fig. 18). Using two one-handed hammers without handles, in 30 hours the miners removed ca. 25 cu ft of trap rock. The mining effort claimed several fingernails from the crew members in the process (William Deephouse, personal communication, 1997). While Thompson's hand-carved birch paddle replica worked well for removing overburden and sand, Bastian reported that a flat piece of tailing was the best way to move rock debris away from the working surface. Bedrock removal was most typically accomplished by crushing rather than fracturing, leaving very small tailings (Bastian 1963:24).

The second stage of mining may have used hammers for more precise scaling and rock removal rather than rock fracturing. These hammers were typically 14–20 cm in length, weighed ca. 2–3 kg when whole, and had a girth of ca. 25–30 cm. They generally exhibited heavy wear, which was perhaps associated with handheld use. In many cases there were multiple superimposed fractures on their surfaces. These smaller hammers were used to remove copper nodules from partially shattered matrix, perhaps with the help of wedges to pry loose rock away from the copper. O'Brien suggested from experimental data at Mount Gabriel, Ireland, that handheld hammers were most useful for precise work in removing trap rock and freeing enclosed metal or ore (O'Brien 1994).

The final stage of hammering was likely situational. Some copper, upon extraction from the vein, held rock inclusions which required removal. This was likely accomplished by pounding and flattening, perhaps on an anvil rock. To do this, the miners probably used small, smooth, oblong cobbles. Small and presumably specialized hammers most likely functioned in a rather "downstream" process of copper manipulation rather than to dislodge copper from living rock. Such hammers may have been associated with copper processing areas similar to the one described by Drier. Baskets, bags, or other containers were probably used to store the freed copper.

The Question of Fire Setting and Quenching

Until relatively recent times when blasting became commonplace, heating through fire setting, followed by rapid cooling through water quenching, was the principal way to reduce the amount of human energy required to quarry rock and mine metals. The heat of the fires weakened the bedrock enough that subsequent sledging with hammers made it possible to loosen and extract the desired materials. Clearly this method was more or less effective depending on a number of factors, including the kind of fuel used, the chemical composition of the target rock, and its physical characteristics.

General similarities in prehistoric mining sites around the world attest to the universality of mining technologies practiced by early peoples (Craddock 1995; Craig and West 1994; Rothenberg and Blanco-Freijeiro 1981; Ryan 1978). For example, in southwest Ireland, Bronze Age miners of ca. 3000 BP removed copper-rich rock and ore from a number of localities. One very well-researched area is known as Mount Gabriel (O'Brien 1994). Using it in comparison to ancient Lake Superior copper mines leads to some new insights about the probable technologies of copper removal. Firing and quenching may have been two such methods.

The classic source for a comprehensive understanding of this process as it was known in Europe during the sixteenth century is, again, the mining work *De Re Metallica* by Georgius Agricola (Hoover and Hoover 1950). But in fact the technique was many thousands of years older than Agricola's century and was used all over the ancient world, first in the quarrying of stone and later in the acquisition of metals and metal ores. Wood or charcoal was stacked against the target rock and ignited, left to burn intensely, and eventually extinguished. Sometimes the process included dashing water on the hot rock, or quenching, to promote sudden contraction and fracturing. It was a process of repetitive burning and quenching, then flaking or pounding away superficially loosened, heat-fractured, or weakened rock. Agricola explained how to apply this pyrotechnology: "As I have said, fire shatters the hardest rocks, but the method of its application is not simple. For if a vein held in the rocks cannot be hewn out because of the hardness or other difficulty, and the drift or tunnel is low, a heap of dried logs is placed against the rock and fired; if the drift or tunnel is high, two heaps are necessary, of which one is placed above the other, and both burn until the fire has consumed them. This force does not generally soften a large portion of the vein, but only some of the surface" ([V]:118–19).

What does a mine in which fires were repeatedly set look like? Some comparative evidence suggested that the face of some fired rock took on a distinctive appearance, a "smoothed wallrock profile which appears to be a characteristic feature of this technique in early metal mines elsewhere" (O'Brien 1994:166). In contrast, hammered walls, according to some authors, "show a fish scale-like pattern of small depressions from battering with stone hammers" (Shimada 1994:50). Other evidence for the use of fire setting consisted of a quantity of charcoal and partially consumed fuel wood deep in the mines. This evidence is strongest in situations in which mine conformation, recovered wood, surface vegetation, and ecological reconstruction allow clear discrimination between accidental and intentional wood deposition and firing.

Focusing on the Lake Superior district, the earliest accounts of prehistoric copper mining routinely cited the presence of ashes, charcoal, and partially burned wood as evidence that fire setting was a necessary part of the ancient mining routine. As Purdy (1982) observed however, the reports assumed rather than demonstrated that fire setting was important. Foster and Whitney commented on the use of fire in regional prehistoric mining to "destroy the cohesion between the copper and the rock" (1850:161) and supposed that mining hard rock without the assistance of fire was impossible for the prehistoric miners. Whittlesey was silent on the topic, but Holmes (1901:690) concluded that the quantities of charcoal he observed at Isle

Royale's mines were evidence of fire setting. Only Fox was a skeptic. He actually tried mining without fire setting and found the results satisfying. His conclusions were based on an assessment of the ancillary labor costs of firing and quenching. Establishing a wood supply and a water source seemed too much added labor for routine use (1911:91).[4]

Bastian's 1962 experiments on Isle Royale bedrock and the effects of fire setting were therefore of great interest. Though they did not reveal the final truth about the techniques of prehistoric mining, they built a body of comparative information from which to make stronger inferential statements about the mining process. His findings through excavation were also of interest, for he concluded that the survey found no direct evidence of the use of fire in the mining process (1963:52). First, in the excavation of Pit 54, a lode mine at Isle Royale's Minong site, virtually no charcoal or ash was recorded. Prior reports of soot deposits were corrected when, upon excavation, the material turned out to be rotted organic remains of rootlets and other soil constituents. Following the earlier experiments of Mills (Ellis 1940), Bastian's firing experiment consisted of setting and maintaining a small fire of intense heat for about two hours' duration. Water was apparently then quickly thrown on the heated rock. Bastian's conclusion was straightforward: "The survey conducted a similar experiment on the Minong amygdaloid, and it had no noticeable effect on the rock" (Bastian 1963:54). Just to confuse matters, however, C. G. Shaw's commentary on the ancient mine he excavated on Isle Royale sometime before 1850 stated that "he found the mine had been worked through the solid rock to the depth of nine feet, the walls being perfectly smooth" (Foster and Whitney 1850:162). Whittlesey, a careful observer of things material, relayed similar findings from the Minesota location (1863:18).

In conclusion, we do not completely understand the role of fire setting in prehistoric mining on Lake Superior. Obviously, bedrock characteristics must be understood before the advantages and costs of firesetting can be understood and calculated. Responses of rock types are highly varied, as are the strengths and effects of individual incidents of firing. Perhaps fissure mines required this kind of treatment. According to the results of experiments elsewhere, composition, grain size, cleavage direction, porosity, absolute heat/duration of heat and draft were some of the important variables to be considered (O'Brien 1994). The role of water also varied, but experimental evidence also suggested that it was not very important. As Winchell suggested about Isle Royale, fuel provision created a different kind of work, not less work. For example, O'Brien's first Mount Gabriel trial used 266 kg of saplings, branches, and brush for a firing experiment lasting two hours. Cutting, collecting, and transporting wood were serious labor concerns, as were means of supplying water. At Isle Royale, water supplies may have imposed further limiting factors (Winchell 1881).

O'Brien pointed out that experiment does not make up for the dearth of technical skills from which modern archaeologists suffer compared to the mining experts of the prehistoric past who, after all, had access to many thousands of years of accumulated knowledge about mining techniques. Modern-day experimental results were highly varied and contradictory. At Mount Gabriel the researchers managed (after setting the 266 kg fire, burning it for two hours, and quenching and pounding for another hour) to remove 12 kg of rock to a depth of 1–2 cm over an area less than a square meter in size. Geological setting, type of mineralization, and lack of experience accounted for most experimental variability (O'Brien 1994:171).

Mining Tools of the Lake Superior District

Hammerstones were the most ubiquitous ancient tool found in association with prehistoric mining ventures. Both mainland and Isle Royale mining localities were littered with whole hammers as well as shattered fragments of waste hammers. Their mining-related contexts, wear patterns, and evidence of intentional modification all revealed the ways that these hammers were used in prehistory. They were a fundamental tool used wherever prehistoric people faced a need to break apart hard objects: at flint quarries, at copper mines, at chert knapping sites, and for food processing needs, among others. They were, worldwide, a universal artifact type, and the similarities in their form derived from similar applications and the laws of physics rather than evidence for worldwide diffusion of technology. In other words, likenesses between hammerstones of widely variable times and places had little to do with culture contact and everything to do with common demands of applying forces in a preindustrial world. The nineteenth-century lake basin visitor Daniel Wilson commented on the gross similarity between Lake Superior hammers and those of the British Isles. After a description of the similarities in technology and material culture, he concluded that

> the resemblance between the primitive Welsh and American mining tools, can be regarded as nothing more than evidences of the corresponding operations of the human mind, when placed under similar circumstances, with the same limited means. It implies an argument, which, if pressed to all its remotest bearings, might rather seem to furnish proof of the unity of the human race, than any direct relations leading to a correspondence in the arts of such widely severed portions of the common family. (1856:229)

The hammers derived from lake and river pebbles of local and glacial origin, with smoothed and symmetrical surfaces from long episodes of water rolling and weathering. They were widely available at active and relict shorelines and riverbeds around the Lake Superior drainage, on Isle Royale as well as on the mainland. The miners routinely made use of 10–15 different kinds of rock for hammers, judging from collections of stones from the Lake Superior region. These hammers had the very important advantage, as far as material evidence goes, of being relatively indestructible, as well as immediately visible archaeologically. Likewise the modifications made to them, through human intention as well as use and wear, constituted a permanent material record of sequent changes and forces to which they were subjected. They were rather expendable as tools go and probably had a relatively short use-life. As soon as a hammerstone's fundamental purpose— to break rock—had been achieved, often through its own breakage, it was discarded and replaced with another. That hard wear is apparent on many of the specimens discussed below; some were riddled with stress fracture lines extending from points of impact deep into the rock. Others were battered at each pole, and most exhibited multiple spalling fractures (shearing of the surface cortex from the poles of the hammer). However, Bastian's experiments suggested that a hammerstone lasted at least thirty working hours and even after that reflected only rather light wear (1963:38).

A rather inelegant and stylistically "mute" tool form compared to their rarer companions, wooden or copper tools, hammers remain relatively unstudied. Fortunately, however, two large collections of prehistoric hammerstones were available for study: one from the Keweenaw Peninsula's Mass City and Rockland area, and the other from Isle Royale. One group of 134 hammerstones and fragments was collected by Roy W. Drier's Michigan College of Mining and Technology Isle Royale campaign of 1953 and 1954. The other collection was recovered from the Caledonia Bluff, Ontonagon County, Michigan, about 1897. These 101 hammers were subsequently acquired, in part by purchase, by the Michigan State Geological Survey. The collection was later maintained at Michigan Technological University's Seaman Mineralogical Museum and the Archaeology Laboratory.[5]

One cannot assume that these 235 examples adequately represent the totality of prehistoric mining hammerstone variability. Any conclusions therefore should not be taken as a last word, but as a first statement about prehistoric mining hammerstones as a cultural entity. Such findings are best identified as hypotheses or speculations to be examined in greater detail when additional comparative collections are available. Not all of the items in the collections were complete, and this affected, to some degree, the total counts of items analyzed. For instance, some were so fragmentary that evidence of intentional alteration was missing, and others were too shattered to collect meaningful measurements. Comparing the degree to which hammers were worn and/or reduced was only possible with whole to nearly whole specimens.

For years the colloquial truth about Lake Superior basin prehistoric hammerstones was that the grooved varieties were found exclusively on the mainland whereas the ungrooved varieties were typical of Isle Royale. While the presence/absence question was settled by Bastian in 1963 (who documented the fact that both kinds were used in both places), the hammers on the island were less frequently grooved than those of some localities on the mainland. But due to the lack of careful and systematic comparisons of collections, insufficient attention was paid to the actual physical evidence that indicated modification. A closer look showed that many hammers were intentionally modified, but not necessarily grooved. For example, a small unprovenienced collection of hammers in the possession of the Michigan Technological University Archaeology Laboratory, believed to be from Isle Royale (Caven P. Clark, personal communication, 1996), included modification evidence on four of seven specimens.

The modification evidence was often rather subtle. Modification included patchy abrasion as well as partial or full grooving. About half of Drier's Isle Royale collection and greater than 90 percent of the Caledonia collection showed evidence of intentional modification for assumed hafting and/or the attachment of a handle. This modification happened in a number of patterns, and it was reasonable to assume that the modes of alteration were chosen in response to a range of other variables, such as intended use, size/weight, style of handle, and strength of material. The basic process of alteration was the roughing and/or grooving of relatively smooth and symmetrical igneous rocks, most of which were water rolled and, to some degree, weathered. This roughing and grooving provided friction or "grab" for handles of roots, sticks, and thongs, which were attached to the hammers in a variety of ways.

In Ireland, O'Brien (1994) suggested that the limited modification observed on the materials from Mount Gabriel was related to the short use-life of the cobbles. Extreme modification was of little importance because the cobbles broke so quickly. It was necessary to question whether some cobbles were equally effective at breaking trap as handheld tools rather than as hafted ones. In addition, some rough grainy minerals, such as the cobbles of Lake Superior, perhaps had surface characteristics that provided enough purchase to anchor withes or binding materials without additional modification, a suggestion made by Holmes (1901). One of our hypotheses then, should question whether modification evidence covaried with the lithology of the cobbles.

There were four types of alteration. *Pecking* of a rock surface to create a friction site represented an alteration style with a very low time investment (fig. 19a). These pecked sites on hammers occurred singly or in groups; they were generally found at the point of maximum edge curvature of hammers. They were not the same as battering or spalling from impact; they were very subtle patchy areas of precise roughening. Rather than establishing a deep groove on a hammer, which in essence is a partial perforation and may weaken the hammer for eventual failure, the hammers were altered just enough to provide purchase or grab for a thong of haft. This spot preparation by grinding or light pecking with a harder rock consisted of a shallow patchy pattern across the margins of the hammers. Spot preparation and alteration of cobbles probably offered a useful strategy for optimizing two potentially opposing needs: haftability versus longest potential use-life. The surface roughing was very shallow; in no case did it extend more deeply than 1–3 mm into the surface of the rock. The makers apparently studied the surfaces and modified them just enough to allow secure withing. Natural features of the rock surfaces were incorporated into the hafting design if useful, or removed by modification if troublesome. These modifications were very subtle, and were in some cases more easily felt than seen, which explains in part why they have rarely been noticed or reported.

Rilling was technically accomplished just as pecking was; slight grinding or pecked indentations along contours provided purchase for hafting (fig. 19b). The rilling patterns, rather than being patchy, were linear. They, again, were very shallow and subtle, and the possibility exists that they represent chafing or erosion of the rock surface from the haft material itself. Rilling represented an increased time investment over pecking for rock preparation. Like pecking, rills sometimes appeared in multiples, and in fact pecking and rilling sometimes occurred on the same hammer.

Partial grooving was an extreme form of rilling and represented an increased time investment in rock preparation, the yield of which was a secure path in which to affix binding material. Instead of simply roughing a pattern, partial grooving resulted in the removal of a visible channel of rock around most of the circumference of the hammer (fig. 20a). The material was probably partially chipped away, then ground smooth through repetitive rubbing of the rock with harder abrasive materials. The resultant channel extended a centimeter or more in depth. The groove channel was generally, but not always, located in the center of the hammer; occasionally it was closer to one end of the hammer. Groove channels also occurred in multiples, both parallel and perpendicular to one another (fig. 20b).

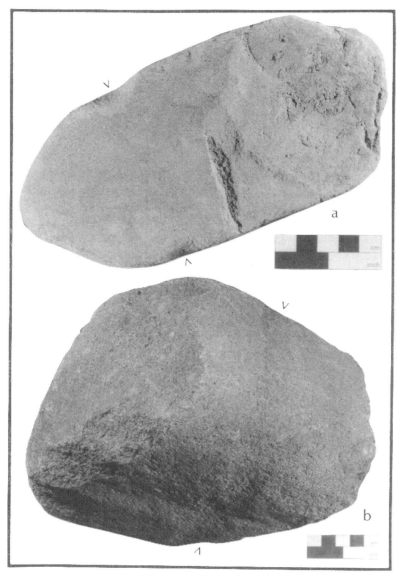

Fig. 19. Intentional alteration of mining hammers: (a) pecked hammer, Isle Royale; (b) rilled hammer, Isle Royale. (Photo by Patrick E. Martin, collections of the Michigan Technological University Archaeology Laboratory.)

Fig. 20. Intentional alteration of mining hammers: (a) partial-grooved hammer, Isle Royale; (b) multiple partial-grooved hammer, Isle Royale. (Photo by Patrick E. Martin, collections of the Michigan Technological University Archaeology Laboratory.)

Full grooved hammers had a groove pattern extending completely around the hammer (fig. 21a). Otherwise they were analogous to partial grooves. They sometimes occurred with partial grooves in the patterns described earlier. Parallel full grooves were reported in some early literature, but none was included in the collection described here. In some cases pecking and rilling patterns accompanied partial and full grooving. The more patterns of alteration and the more surface area dedicated to them, the greater the probable time invested in the preparation of a hammer, the greater its assumed specificity of function, and the more effectively it may have functioned, both in terms of useful life and reduced maintenance and repair.

The styles of alteration on prehistoric hammers in the Lake Superior district ranged from "none" to those with "full/partial grooves" (table 2). The category "none" represented an expeditious strategy for mining: no rock preparation and immediate discard upon fracture beyond further use. No alteration at all suggested handheld use or the use of a thong or haft that operated adequately with no obvious surface alterations to the rock. Logic suggested that handheld hammering might have produced hammers with heavy wear on one use pole and little use of the other pole.

In table 2, the modification methods were ranked in levels of assumed decreasing intensity and time requirements, from multiple grooving to no surface alteration. The Isle Royale sample showed a distinct pattern of minimal alteration. At Isle Royale, expediency of preparation and use appeared to be the pattern among the hammers analyzed. At Caledonia, a less coherent pattern emerged, but one which favored the more intensive modes of alteration.

Some of the less obvious surface alteration may have been produced by scouring or wear from the inclusion of tightening shims into the withing material, a technology that was reported from prehistoric Chilean copper mining sites and probably represented an essential technique of withing. Craddock (1990) reported that shimming was constantly needed when adjusting the tension of withing, which loosened rapidly during her experiments with replicated prehistoric mining tools. Its co-occurrence with partial grooving was interesting in this regard; the mechanics of use may have produced the need for shimming in hammers that were insufficiently grooved. In fact, haft failure was more troublesome than hammer fracture during the use of experimental stone hammers (O'Brien 1994:170). Holmes made reference to hafting polish reported on Isle Royale hammerstones (1901:694), which he stated encircled some hammers at the point at which a haft or withe would logically have occurred, but Bastian discounted such evidence (1963:35). Such polish was not visible on Isle Royale hammers in the collection examined above, but roughing was apparent.

Fig. 21. Variability in mining hammers: (a) full-grooved hammer, Portage Lake Ship Canal, Hancock, Michigan; (b) small hammer for fine crushing, Isle Royale. (Photo by Patrick E. Martin, collections of the Michigan Technological University Archaeology Laboratory.)

Table 2. Intentional modification by frequency, locality, and rank order: Isle Royale and Caledonia hammerstones.*

Mod Type	f - Isle Royale	Rank order	f - Caledonia	Rank order
Full Groove/Partial Groove	0	—	2	5
Full Groove	6	6	57	1
Partial Groove (PG) (mult)	0	—	2	5
PG/Rilling	0	—	0	—
PG/Pecking	5	7	5	4
Partial Groove	7	5	24	2
Rilling/Pecking (mult)	2	8	1	6
Rilling (mult)	7	5	0	—
Rilling	10	4	1	6
Pecking (mult)	14	3	2	5
Pecking	15	2	0	—
None	41	1	6	3
Totals	107		100	

*Frequencies (f) reduced by eliminating some broken specimens.

Differences in mining hammers from Isle Royale and mainland mining areas implied the operation of a set of selection criteria for a particular lithology, size, shape, and weight range (fig. 21b). There also appeared to be some shape distinction between hafted cobbles and unhafted ones, with the hafted cobbles representing a narrow range of shape variation. Overall, the Isle Royale collection of hammerstones contained a broad range of extrusive volcanics derived locally and from glacial till. At least fourteen rock types made up the collection. All were water rolled, smoothed, and symmetrical. Basalt was the most common raw material, making up nearly half of the collection. About half of the hammers exhibited subtle to formal intentional surface alterations that provided a seat and purchase for hafting and or handle attachment. All phases of the use-life of a hammerstone were represented, but most were lightly to moderately worn on two poles. The range of intentional modification patterns appeared to co-vary, to some extent, with material and wear. The collection probably contained some hammers that were used in specific ways, but the sample was too small to determine how certain hammers and hafts may have functioned.

Caledonia hammers were more homogenous in material and surface modification than those from Isle Royale. Most of the specimens were basalt, with gabbro and diabase well represented. The range of raw materials was narrower than that seen in the Isle Royale collection; the top three rock types accounted for greater than 82 percent of the collection. Partial to full grooving was the most frequently noticed pattern of modification. Most hammers exhibited wear on both working ends, but overall the wear appeared less intense than on those hammers from Isle Royale. The Caledonia hammers exhibited the entire range of modification alternatives. Though there was no apparent link between hafting modification and the weight of a hammer, the single extremely heavy hammer in the Caledonia appeared to have complex modifications for hafting.

The patterns of fragment spalling from the hammers indicated a great deal about the amount of force expended on their working surfaces as well as the direction of the forces applied. Frequently cracks or partial fracture lines marking the path of forces proceeding through the object were visible on the surface of these tools. It is clear that rocks of a greater strength and hardness than the local trap material probably had considerably longer use-lives than those of the same or less hard materials. One presumes then that local characteristics of trap exercised some control over the lithologies of the rocks used for mining.

The people on the mainland used a limited range of rocks and had a curated technology, putting a lot of work and time into preparing hammers for use. The people on Isle Royale used a relatively broader range of rock types and practiced expedient modifications upon them, doing as little as necessary to alter and prepare them for use. Highly modified hammers on Isle Royale were always of a particular rock type: gabbro. One hypothesis about this finding is that these altered gabbro hammers were brought there by people from elsewhere, likely the south shore, where gabbro hammers with high modification were more common. This speculation was advanced by Winchell many years ago (1881:605), but it is interesting to find that it is a possibility backed up by archaeological evidence. Some recent research indicated that during prehistory the island was used by whomever wanted to go there (Clark 1991, 1995). Hypothetically the difference in the hammers might be attributable to differences in access to territory; the mainland may have been the traditional territory of a limited social group.[6] Taking a sample of island hammers and comparing it to a sample of mainland hammers, one finds that the pattern of intentional alterations in each sample is reversed. This in fact would fit very neatly with other models of Isle Royale's use: open to many groups, short-term stays, expedient technologies with an emphasis on ad hoc local materials, and multiple episodes of use. Caledonia suggests the opposite: repeated and fairly exclusive use by a consistent (perhaps local) group of people, reduced variety in technology and material, and a strategy of curated rather than expedient tool production and use.

Conjectures about local hafting styles, materials, mechanics, and technologies are necessary because only rarely were such data recovered in solid archaeological context. One must rely on analogies with similar materials recovered archaeologically or ethnographically from groups of people exhibiting similar adaptations to similar environments, attempting to accomplish similar tasks and functions. Fortunately, some analogous objects were recovered under well-documented conditions from other prehistoric mining localities around the world. These are included as analogies to the probable means by which the North American miners arranged their hafting and withing materials. These constitute some of the best data available on the hafting of prehistoric mining tools, and together with additional insights from archaeology, provide a least an outline of how hafting of stone hammers for mining and other tasks probably took place. The technology is found virtually worldwide and has no known specific origin point.

Experimental studies based on withes recovered archaeologically from Mount Gabriel suggested several alternative ways of attaching hammers to handles or withes (O'Brien 1994:130). Rigid handles likely absorbed forces and broke easily as well as transferred force to the wrist rather than to the hammer. Flexible withes or bindings would have been most practical in transferring increased energy through momentum to the hammer itself, which would have allowed the delivery of a larger force to the target rock. Withes were designed to maximize force when rock faces were the targets of blows (134), but other kinds of chores required specific designs; rigid handles were appropriate for crushing smaller pieces of rock and freeing metal (or ores) from it. Interestingly, precision work with hammers was most readily accomplished when they were handheld rather than hafted (170). At Mount Gabriel, flexible hazel "withy" was wrapped around the hammer and embedded into the groove, then secured tightly by weaving and/or tying off the ends. Rigid handles were also known; the opposing pieces of the handles followed the groove and were then wrapped tightly together to provide a grip. At other early mining sites in continental Europe, such as the Rudna Glava (Yugoslavia) mines and those in Spain, large hammers with central grooves, discarded after use, were also commonplace. These simple tools were widely distributed wherever ancient mining was found, though they took on specialized shapes (and presumably, functions) from one place to another (Jovanovic 1978:340; Rothenberg and Blanco-Freijeiro 1981).

Similar hammers have been recorded from many localities in Central and South America (Craig and West 1994:13–14, 308) at silver, tin, and copper mines. At Huantajaya, Peru, a hafted hammer recovered in a silver mine consisted of a grooved hammerstone bound between two rigid handle portions by leather or hide strips. Hafted hammers associated with wedges for shimming were reported by West from Mexico (13). Andean copper miners used hammers of local rock, both hafted and unhafted, of a range of weights, which were minimally modified and battered from use (Bird 1979:114).

In the Lake Superior mining district, direct evidence of hafting modes was rare. Early reports from Isle Royale included mention of the recovery of knotted rawhide (Winchell 1881:603-4)—presumably hafting material—within the mines. Fox's informants had seen both wood and rawhide in the ancient mining pits, but argued about whether there were grooved hammers on the island (Fox 1911). Whittlesey reported the discovery of a hafted grooved hammer at the Cliff Mine, which specimen "was found with a root of cedar still twisted in the groove, but so much decayed that it fell to pieces and was not brought away" (1863:10).

The remains of a possible ancient withe were recovered from the Delaware Mine vicinity of Keweenaw County, Michigan, in 1994. This badly decayed possible withing consisted of a root or vine, woven in a distinct pattern around two perpendicular twigs or remains of wooden handles. It was reportedly found while several amateur archaeologists excavated in the vicinity of the Delaware Mine, Keweenaw County, Michigan, and was associated with a number of hammerstones and hammer fragments. Because no one recorded its condition and circumstances of recovery in a way that can be verified, its authenticity or potential associations with other materials of prehistoric age is uncertain.

Other Artifacts Associated with Prehistoric Mines

The accounts of the earliest investigations of prehistoric mining reported that paddle-shaped wooden objects, thought to have functioned as shovels, were occasionally found within the ancient mining pits. Such items were reported from the Waterbury Mine in Keweenaw County, and were identified as white cedar. Their blades were much reduced by wear "more worn on the under side than the upper, as if the mineral had been scraped together and then shovelled out" (Whittlesey 1863:8). Similar tools were found at the Owl Creek vein near the Copper Falls mining location (10) and at McCargoe's Cove at Isle Royale (Winchell 1881:603). The shovel at Isle Royale was described as one foot long and ca. 4–5 inches wide, with a broken handle. It was asymmetrically worn "which showed by its worn and battered side that it had been used in moving dirt" (603).

Numerous historic sources reported the discoveries of wooden bowls or containers associated with mining pits, though no examples were recovered from controlled archaeological collections. Bowls may have served as water bailing devices, and/or collecting pans for recovered copper fragments. Describing materials from McCargoe's Cove on Isle Royale, Gillman recorded that a large portion of a wooden utensil, shaped like a bowl, was taken from among the debris, charcoal, etc., at the bottom of a pit. This vessel had possibly been used in bailing water from the excavation. It must have originally been about three feet in diameter, and something of the rude character of the tool employed in shaping it could be gathered from its appearance. It was not of uniform thickness throughout, the wood having been more easily removed when working in certain directions; e.g., when cutting with the grain, the vessel was thinner in these portions. (1873:173)

Whittlesey reported that a wooden scoop or bowl was discovered at the Waterbury Mine on the mainland "evidently intended to dig up and to pass water. Its edge had been worn, like the shovels, by scraping over the rock; but it was so much decayed that it fell to pieces when it was taken out" (1863:8). A similar item was reportedly found at the bottom of ancient workings near the Forest Mine, Ontonagon County, Michigan, "which, from the splintry pieces of rock and gravel imbedded in its rim, must have been employed in bailing water" (Foster and Whitney 1850:161).

Copper chisels, gads, pikes, and other items, sometimes of wood, were assumed to have functioned as prying implements or wedges to loosen copper from bedrock. As was the case with other tool categories, the evidence for the association of these tools with prehistoric mining activity lay in the written accounts of early historic reporters rather than in evidence derived from controlled archaeological excavation. The early chroniclers sometimes connected the wooden remains with mining because they noted that the wood had been trimmed with metal tools.

Foster and Whitney, in their accounts of the original discovery of ancient mining near the Minesota Mine on the south shore of Lake Superior, related that "a copper gad, with the head much battered, and a copper chisel, with a socket for the reception of a handle were brought to light" (161). Copper tools were reportedly found in association with mining activity at the Phoenix Mine in Keweenaw County (162) and on Isle Royale (Winchell 1881:602). Stone hammers were discovered with fracture patterns on which "the peculiar sharp cut character of the fracture in many cases indicates that the implement had been used to drive metallic wedges, such as quarrymen call a 'gad' " (Whittlesey 1863:11). Wooden levers were reportedly noticed at the Minesota Mine, preserved by waterlogging. Other wooden elements were reportedly discovered with these remains, notably, a ladder, carefully trimmed "so as to leave the stumps of the branches projecting, on which men could readily descend or ascend to or from their work" (19). Both Whittlesey and Winchell reported the discovery of wooden timbers within mines, which had apparently been trimmed and shaped with metal tools

(Whittlesey 1863:7; Winchell 1881:603). The functions of these so-called timbers, if they were purposefully used in mining, is not clearly understood, but Holmes suggested that they were used to clear away loose and weathered rock. "Wooden implements would have served to loosen and remove the superficial materials, laying bare the rock surfaces and exposing protruding masses of copper" (1901:692). Wood (1907), looking carefully at the incisions of tools on wood from a mining pit, determined that they had likely been made with narrow copper tools, and were, according to him, intentionally placed in the pit.

Early accounts occasionally reported structural remains related to mining activities, both in the Ontonagon district and on the Keweenaw Peninsula. Wood reported the discovery of white-birch pole scaffolding at a depth of ca. 7 feet within an ancient pit near the Quincy mining location, "which had been used as a landing, to assist in removing the excavated rock" (288). At the Minesota Mine the excavation and discovery of a large mass of native copper supported by the systematic placement of billets and sleepers made up the initial spectacular evidence for prehistoric mining. Whittlesey later described the mass and its wooden support in detail: "It lay upon a cob work of round logs or skids six to eight inches in diameter, the ends of which showed plainly the strokes of a small axe or cutting tool about 2½ inches wide . . . the mass of copper had been raised several feet along the foot wall of the lode, on the timbers, by means of wedges" (1863:18). Whittlesey also reported the discovery of what he termed "gutters," for water drainage, at the Waterbury Mine locality on the Keweenaw Peninsula: "Beneath the surface rubbish the remains of a gutter or trough composed of cedar bark were discovered, the object of which was clearly to conduct off the water which was bailed from the mines by wooden bowls" (7). The shapes and functions of such remains were not verified archaeologically.

Holmes (1901) and Bastian (1963) reported the discovery of flaked stone objects of unknown functions associated with mining activity: "Among the stone implements found are a few forms that may properly be classed with the flaking hammers of the flint quarries. All are small and somewhat discoidal, and are flaked and battered more or less completely all around the periphery" (Holmes 1901: 694–5). Holmes proposed that they were used to produce other stone tools, or that they were used to groove the hammerstones. But based on observed wear patterns such as glassy surface polish, Bastian concluded that such patterns suggested a scraping function. There were no reports of discoids from the mining areas on the south shore of the lake nor on the Keweenaw. These artifacts were considered difficult to recognize in the field, their subtle use-wear patterns being their most distinguishing characteristic (Bastian 1963:34). Their function related to mining remains conjectural.

Most historical reports of mining activity commented at length on the frequent occurrences of well-preserved organic materials and bark, soot, charcoal, and partially burned wood within the mining detritus and in the pits themselves. However, distinguishing between accidental and purposeful deposition of them was difficult. Without irrefutable evidence of cultural causes, natural processes of deposition (such as forest fires) were assumed to be responsible for these remains. Though it is known that aboriginal people in the region used bark profusely in all kinds of constructions, from baskets to houses, no confirmed examples of intentionally worked bark containers have come from the mining pits. Charcoal and ashes were equally problematic. There was nothing particularly distinctive about the burned wood in the pits in terms of species, shape, and other characteristics that distinguished it from background vegetation, apart from occasional claims that partially burned or partially preserved wood showed scars indicative of intentional cutting. These materials were a source for radiocarbon dates if derived from known contexts with depositional histories relating them to episodes of human activity.

Conclusion

Systematic searches for the material evidence of mining began in the Lake Superior region about 1847. A review of the mining evidence from excavation records, historical accounts, and cross-cultural analogies suggested simple mining technologies that remained rather stable through prehistoric times. There was no major technological change through time in mining methods evidenced by the prehistoric archaeological data. Some variability likely occurred as a response to local and particular conditions, such as the availability of useful hammerstone material, the shape of a specific copper vein, and the tenacity of the local bedrock. Some people may have resorted to fire setting, but its extent and actual utility is not yet well understood.

The mining people were keenly aware of their environmental surroundings and how to identify their constituent resources. Copper deposition was one more element in that environment whose characteristics of deposition were hardly mysterious or difficult to predict. Sufficient quantities of copper were recovered in the course of ordinary pursuits, such as food gathering and hunting or fishing, to supply most needs (Clark 1991). There was no archaeological evidence that large numbers of people ever mounted stupendous expeditions or voyages for the sole purpose of acquiring copper.

The archaeological data related to mining indicated that small numbers of people repeatedly visited reliable localities and took away limited amounts of copper. In the case of Isle Royale, the evidence suggested that people from a range of small local hunter-gatherer groups occasionally traveled to the island to camp and gather fish and copper. In contrast there may have been some greater level of exclusive use of the Caledonia region by a resident family group or groups. The actual nature of the social groups that mined and the degree to which they may have controlled access to a particular copper source, is not well understood. These kinds of data can only be derived from careful excavation and meticulous comparisons of sets of data from multiple sites, and this work has not been done in the region on a wide enough basis for satisfactory conclusions.

There is much to be learned, however, from extant collections of artifacts from the mining sites. Hammerstones, for instance, are not mute rocks but artifacts of wide human use whose fracture patterns have revealed much about their functions and use-lives. The widely varied intentional alterations on hammers from extant collections are valuable data, and studying them makes it possible to understand more fully how the tools were actually used. In addition it is possible to reconstruct how and why certain choices were made about hammer material and hafting designs, and why certain wear patterns co-occur with materials and hafts. Instead of the rather dead-end conclusion that grooved hammers occur more frequently on the mainland, we now understand that the nature of local materials, the need for impromptu tools, the characteristics of the bedrock at hand, and materials at hand for hafting may have tempered decisions about how to alter and use a hammer.

Despite the fact that the prehistoric data have been ravaged by site destruction, casual collecting, and the developments of industrial-scale mining, there are profitable comparisons to be made with other parts of the world where initial metal mining occurred. Studying mining hammers from Mount Gabriel, Ireland, helped to establish hypotheses for understanding their variable uses on Isle Royale and at Caledonia. Hafting technologies from Mexico and Chile stand as analogies to those which may have been used in north central America. Fire setting experiments from other places offer instruction on how fire may have been used in local mining efforts.

Suggestions for Further Reading

Craddock, Paul T. *Early Metal Mining and Production.* Washington, D.C.: Smithsonian Institution Press, 1995. Worldwide comparisons of initial mining exploits known from archaeological evidence.

Craig, Alan K., and Robert C. West, eds. *In Quest of Mineral Wealth: Aboriginal and Colonial Mining and Metallurgy in Spanish America.* Geosciences and Man Series 33. Baton Rouge: Louisiana State University Press, 1994. A gathering of the writings of many experts on archaeological and historical evidence for New World mining and treatment of ores.

Hoover, Herbert C., and Lou Henry Hoover, eds. *Georgius Agricola: De Re Metallica.* Translated from the first Latin edition of 1556. New York: Dover Press, 1950. The classic account of medieval European mining, written down for the first time.

O'Brien, William. *Mount Gabriel: Bronze Age Mining in Ireland.* Galway: Galway University Press, 1994. A comprehensive site report about a well-preserved copper mine complex.

5
Implements, Ornaments, and Objects of Faith of Great Variety
—William H. Holmes, 1919

Ancient Technological Practices

> I have neither seen nor heard of a single object of copper from the mounds which I cannot reproduce from native or nodular copper with only primitive appliances of the kinds described, by successive processes of stone-hammering, beating and rolling, scouring, embossing and grinding.
> —Frank H. Cushing, 1894

Introduction

Thus declared Frank Hamilton Cushing over one hundred years ago. Though he was not the only researcher to try such replications (Willoughby 1903), he was the most convincing of late nineteenth-century specialists to demonstrate that the elegant copper foil work found in association with prehistoric mounds was within the technological repertoire of native metalworkers. The debate about metalworking origins raged for decades and refused to die away; it was raised again at the 1892 meeting of the Anthropological Society of Washington after Warren K. Moorehead's presentation. Moorehead's discussion of the artistry of the metalwork discovered at the Hopewell site in Ohio provoked a number of listening scholars. They concluded that such carefully executed and complicated pieces of copper work were "perhaps of European manufacture, or granting the art-work on them to have been native, that the copper plates from which they had been cut must have been of foreign make, since such large thin sheets of metal could only have been wrought by means of roller mills or stamping machines of hard metal" (Cushing 1894:93).

Cushing countered with an 1894 response published in the *American Anthropologist,* in which he described the how-to's of working unsmelted metals, based on the observations he made among Zuni silversmiths of the American southwest. Although relatively recent Zuni silver working may have in part been derived from contact with European smiths (Leader 1988), Cushing's point was not to show that the Zuni originated their native metalworking in isolation of anyone else. Rather, he wished to demonstrate that basic copper-working technology was within the reach of prehistoric peoples, using their conventional tool kits. Cushing demonstrated this likelihood by using what he termed "stone age appliances." He replicated the embossed and cut-out forms of copper foil work found at the Hopewell site and thoroughly described the probable methods for their fabrication. These abilities, he reasoned, were developed on other materials before they were transferred to metals. "Thus I am led, by the experiments related below and by other considerations, to suppose that the simpler of the aboriginal arts in metal were at first influenced by more than one antecedent art, namely, not only by various methods of stone working, but also of bark-working, skin working, horn-working etc." (Cushing 1894:97). Cushing went on to show how these accumulated methods could have been applied to a new material, copper, by the people of riverine central North America during 400 B.C.– 400 A.D. His efforts introduce two major themes of past research that deserve emphasis as we look now at the outcomes of mining, the artifacts produced by native metalworking, and their transformation into cultural materials.

One theme is the experimental replication of artifacts. As early as the late 1870s, avocationalists in Wisconsin attempted to replicate the prehistoric copper artifacts discovered in the eastern parts of the state (Draper 1908; Hoy 1908). Following Cushing, a sizable list of research specialists added to our knowledge about how to work metals with "primitive appliances" (Childs 1994; Clark and Purdy 1982; Franklin 1982; Leader 1988; McPherron 1967; Schroeder and Ruhl 1968; Smith 1965; Steinbring 1975; Tylecote 1992; Willoughby 1903). Topics of particular interest include an understanding of hammering, particularly the production of thin sheets (Clark and Purdy 1982); the effects of heat treating; hardening techniques; molding to create complex forms; and final finishing techniques. This chapter will introduce some of the techniques of prehistoric metalworking and highlight some of the results of systematic experimentation with them.

Experimentation is only one avenue toward understanding what processes were actually practiced in prehistoric metalworking. Some investigators developed an understanding of the physical and behavioral changes metals take on as result of manipulation by physical forces such as hammering, heating, cooling, smelting, and casting. The microstructure of an artifact is in part a physical record of the forces to which that artifact was subjected (Clark and Purdy 1982; Vernon 1990). This record allows the careful researcher to trace the history of the artifact's physical treatment and the processes accompanying its fabrication. A comparison of experimental versus actual artifacts on a microstructural level suggests that specific sequences of experimental procedures could account for artifact production. Comparative examination of artifacts and raw materials from a range of time periods allows the reconstruction of likely sequences of technological innovation that may have taken place during prehistory (Leader 1988). Unworked metals, studied in an "as received" condition straight from known geological contexts, provide a baseline for understanding the probable treatments materials receive as they become cultural products (Stevens 1996; Sutter 1993). Such work is practiced worldwide by dozens of scholars who term their work archaeometallurgy, or the systematic study of the production and use of ancient metals to understand their physical and chemical properties, their significance to cultural institutions, and the ideology and symbolism involved in both their manufacture and use. A major goal of this chapter is a review of the successes of archaeometallurgy in honing an understanding of prehistoric metals in their social contexts (Craddock 1995; Glumac 1991; Maddin, Wheeler, and Muhly 1980; Tylecote 1992).

Both metallographic examination and experimentation are vital toward an understanding of prehistoric copper-working technologies. Fortunately a number of researchers worldwide have collected a large body of data on this subject. Smith (1965) suggested, as have others (Vernon 1990), that the technical history of a copper artifact was embedded in its structure and a study of metallography (the examination and description of the physical appearance and microstructures of cross-sections of a metal object) told this story far better than anything concerning its metallic composition. There are many laboratory methods to examine the microstructures and composition profiles of archaeological metals.

In a typical metallographic examination, archaeological and control specimens of metals are carefully cut in cross-section, mounted, ground, polished, and etched before they are examined under magnification and photographed (Vernon 1990:500). The examiners pay particular attention to microstructural comparisons of grain or crystal shape and size, as well as the appearances, positions, and characteristics of grain boundaries. The latter vary systematically depending in part on stresses, such as hammering and heat, to which an object has been subjected. Heating to critical temperatures, or *annealing*, institutes a process of recrystallization, whereby metal grains slowly approximate their prehammered state. Recrystallization is important to metalworking because cold hammered metal loses its malleability and becomes brittle as work proceeds. Annealing basically removes strain in the metal, which builds up through cold working, and restores malleability. The recrystallization process may be accelerated by slow heating and cooling. Metallographic examinations can investigate questions of intentional hardening related to intended functions of cutting edges and the degree to which objects were prepared for efficient work, as well as whether the raw metal was first cold hammered, smelted, or melted and cast.

Other common goals of metallurgical analyses of ancient copper are to assess a sample's composition and the presence/absence and quanta of trace elements and added metallic elements for alloying. In such studies, an array of analytical techniques can be applied, dependent upon desired levels of accuracy and the particular suite of elements to be detected. Common analytical processes to determine the chemistry of an object include neutron activation analysis, atomic absorption, and x-ray fluorescence. Levine (1996:9) summarized current analytical approaches and their utilities, particularly for the detection and measurement of trace elements in native copper samples.

Studying Prehistoric Copper Working by Microscope and Experiment

Inspired by the same set of events as Cushing, Willoughby (1903) published a general account of his experiments with prehistoric metalworking. For the sake of consistency, his terminology guides this review of how ancient people formed copper artifacts. Willoughby's procedures identified seven basic steps in the probable construction of Hopewell-era copper earspools. While such items do not represent the most typical prehistoric artifacts, the seven steps to make an earspool include all of the general metalworking tasks that prehistoric people routinely used to fashion most forged copper artifacts. These were hammering, annealing, grinding, cutting, embossing, perforating, and polishing. This order intentionally represents increasing complexity and specialization as well as the logical sequence of steps from an artifact's start to finish. Other less common techniques also deserve attention: stretch hammering or "fabricating thin sheets from lumps of metal" (Clark and Purdy 1982:45); joining, folding, and riveting (Childs 1994; Franklin 1982); swedging and sinking (Draper 1908; Hoy 1908; Leader 1988); and pressure welding, hot forging, or bonding (Childs 1994; Leader 1988; Wayman, King, and Craddock 1992). There is still considerable debate and some uncertainty among metalworking researchers about the degree to which some of these specialized techniques were present in prehistoric fabrication practices.

Cold hammering is the fundamental process that accounts for the shape, strength, and hardness of prehistoric artifacts made of native copper. Copper is a crystalline solid at temperatures less than 1083 C, which can be shaped or deformed by applying pressure through hammering. During this process individual grains or crystals of copper, which at first are irregular in size and shape, are altered in somewhat predictable ways. Copper is a relatively soft and ductile material in its native state, so it tends to compress and spread when force is applied. The result is compaction and reduction in original thickness, deformation of grain structure, and the generation of heat. The strain of deformation accumulated by copper in its initial reduction in thickness is relatively great, and malleability or workability is lost quite quickly in the hammering process. Vernon (1990) identified six general stages in grain deformation that characterize prehistoric copper working (table 3). The degree to which these characteristic structures occur in a specimen of worked copper is a rough but useful measure of the most recent cold-working or hammering event to which it was subjected. The speed of recrystallization is reduced in specimens that are heavily cold working.

Shaping copper by stone hammer required a long series of delicate, well-positioned blows to shape and reshape it into its many final forms. Cushing suggested that this technology transferred logically from stone working, that basic human skill of greater than two million years' duration. "It is safe to assume, as a general proposition, that no new art was ever practiced by aboriginal Americans as strictly new. No art, I mean, in the working of new or unaccustomed material, which was wholly uninfluenced by arts and methods, which, in connection with other materials more or less like the new material, had been practiced before" (Cushing 1894:97). Cushing linked metalworking with well-established technologies in stone, wood, hide, shell, and bark and suggested that it was inconceivable that the same people who quarried flint did not also discover the malleability of metals.[1] Maddin and colleagues (1980) made essentially the same claim for the earliest metalworkers in southwest Asia and adjacent Europe.

Table 3. Stages of copper deformation during cold hammering (after Vernon 1990:503).

Stage	Characteristics	Notes
Initial	Grains relatively undeformed	No cold-working apparent; grains heterogeneous
Moderate	Elongation of grains	Appearance is of unidirectional structure
Intermediate	Near-absence of original grain boundaries	
Advanced	Some flow lines	Crystalline structure no longer apparent
Penultimate	Absence of grains, consists of flow lines and strain markings	
Ultimate	Consists entirely of flow lines	

The earliest replicative experiments of native copper working in the upper Midwest were conducted by Hoy around 1879. He recreated a copper ax through the process of cold hammering and bequeathed it to the State Historical Society of Wisconsin. But he suggested several unusual twists to the hammering process, ones that have never been substantiated as used in prehistory. The ax, he claimed, had been swedged, or hammered into a prepared form to replicate its shape. Hoy also speculated that copper artifacts were sometimes rolled between two flat rocks to effect their cylindrical forms, "which is the manner in which several of the articles in the collection of the State Historical Society might have been made" (Hoy 1908:172). He replicated these implements experimentally using his rolling method and deposited them with the museum. "I have cylindrical implements tapering regularly from the center to the points, as well as the beautiful hatchet referred to, made to illustrate in evidence of my position" (173). Clearly, there was more to be learned by experimentation. Hoy's experiments did not substantiate the actual use of such techniques in prehistory, but only their possibility.

Both Cushing (1894) and Willoughby (1903) experimentally replicated rather elaborate sheet copper artifacts. They concentrated first on hammering the copper to thicknesses comparable to those of Ohio Hopewell artifacts such as earspools and embossed cut-outs. Both used tools that were widely demonstrated to have been part of aboriginal tool kits. In fact Cushing remarked that he could not achieve the desired similarities to native articles without the exclusive use of native tools (1894:96). This was an important finding, because Cushing and Willoughby wanted to counter the claims of some prehistorians that native metalworking had derived from European materials and technologies. Willoughby restricted his tool kit to objects he found or made on the very beach where he conducted his experiments: flat lake rocks, pebbles, a bone, a flint, and a small fire. He began with two nuggets, one of Lake Superior copper and the other from the Ohio mounds. Willoughby concluded that both careful hammering and repeated heating were required to form a sheet of copper of a desired thinness.

Other researchers later discovered that it was possible to reduce copper by as much as 90–96 percent of original thickness simply by careful cold hammering (Frank 1951; Schroeder and Ruhl 1968; Smith 1965), though it is not clear whether they relied exclusively on replicas of a prehistoric tool kit during their experiments. The most extensive replicative research concentrating on hammering, at least from an archaeological perspective, was done by Clark and Purdy (1982) and Leader (1988). Clark and Purdy were particularly interested in a process they called stretch hammering:

"[B]eginning with a lump of native copper, a large thin sheet of fairly uniform thickness could be fabricated using a smooth river rock as a hammer" (1982:46). In other words, this is the same process that Cushing and Willoughby attempted to re-create.

Clark and Purdy further determined that the degree to which cold working alone could reduce initial thicknesses to fine sheets without splitting or failure was affected by the presence of impurities within a particular specimen (47). They reported that the natural heat increase that occurred as copper was cold hammered was not high enough to begin to recrystallize their specimens. Because the actual temperature at which recrystallization began was in part a function of the relative purity of the specimen, they concluded that this temperature may vary by several hundred degrees (53). The purer the copper, the lower the activation energy for recrystallization. This finding suggested that prehistoric copper workers were probably aware of these variable properties, which could have affected the treatments to which an artifact was subjected. The geological sources of copper that were preferentially exploited may have varied depending on their purity (56).

Leader's research (1988) focused in part on experimental replication of copper-working techniques and their changes through time. He claimed that the increasing skill in hammering technologies between ca. 500 B.C. and ca. 1500 A.D. related to the growth and spread of specialist craft production in the riverine southeast of prehistoric America. Most of his study specimens came from cultures representing relatively recent manifestations of prehistoric copper working rather than ancient ones. Despite this, "Archaic copper artifacts are functional and demonstrate understanding of the more difficult manual techniques of metal working" (Leader 1988:46). He noted the presence of "hammer packing" or "localized hammering to increase metal density" (65), a technique that allowed a copper implement to maintain a tougher working edge. Sockets of Archaic-period tools were carefully hammered, perhaps around preformed blocks (mandrels) of wood or other material, to assist in their shaping. At least during the Middle Woodland period, specimens were tested by hammering for potential flaws and inclusions. Appropriate forming technologies were in part dictated by the size and purity of the specimen (78–79). Wayman, commenting on the appearance of hammered native copper from a metallographic perspective, came to these supporting conclusions: "The hammering technique, although primitive in itself, was carried out with assurance and skill; the sheet metal is often of fairly uniform thickness even though laminations are sometimes present" (Wayman, King, and Craddock 1992:133–34).

The Role of Annealing

The necessity of annealing and the evidence suggesting its wide application in prehistoric copper working has been considered by virtually every experimental archaeologist and archaeometallurgist (Childs 1994; Clark and Purdy 1982; Leader 1988; Vernon 1990; Willoughby 1903). The characteristic changes in copper that develop as a result of repeated hammering and annealing also have been studied over the past sixty or more years by a number of materials scientists who have applied their expertise to analyses of prehistoric copper artifacts (Coghlan 1951, 1960; Frank 1951; Franklin 1982; Miles 1951, Root 1961; Schroeder and Ruhl 1968; Smith 1965; Wilson and Sayre 1935). Raising the temperature of relatively pure native copper above 225 C initiates recrystallization, which restores some of the grain deformation produced by hammering. In this process, some of the malleability of copper is regained, but the hardness of the object is reduced drastically. Every archaeometallurgist reports that to some extent or another, annealing was routinely and repeatedly practiced in prehistoric metalworking. Re-establishing the actual sequences of hammering and annealing episodes on a specific artifact is not a simple matter, because the evidence of the sequence is wiped out by repeated processing events. Obviously the sacrificing or altering of rare metal artifacts in order to gather data for such studies also imposes a constraint. Finally, there is a sampling difficulty as well, because treatments may vary over the surface of an individual artifact.

There is at least one long-standing question that annealing data pose for archaeologists. That is centered on the fact that some prehistoric artifacts, thought to have been used as implements for doing work tasks such as cutting, scraping, and piercing, were shaped to their final form and then apparently annealed, leaving them relatively soft rather than hardened by final hammering. The implication is that this soft state interfered with an implement's functional efficiency. Wayman, however, noted that natural processes of recrystallization, as well as circumstances such as cremation or forest fires, could reproduce in an artifact the same metallography as intentional annealing might (Wayman 1985).

The role of annealing in replicating prehistoric copper artifacts was examined early on by Willoughby, who found that repeated cycles of hammering and annealing were required to form copper into sheets of appropriate thickness and workability: "[A] few blows sufficed to show the tendency of the copper to crack along the edges as it expanded. This tendency was overcome by annealing" (1903:55). As the production of the replica artifacts proceeded, the annealing process was repeated as needed. It was particularly critical when the maker desired the copper artifact-in-the-making to bend smoothly. Frank (1951:59) stated that 90 percent thickness reduction was the maximum that could be achieved between anneals, but it was not clear whether he derived these estimates from experimental stone hammering or from other sources. Frank's metallurgical examinations suggested that although the artifacts he studied had likely been treated to temperatures above 225 C, there was no evidence of lengthy anneals. He concluded that the artifacts could not have been fabricated without repeated heating for recrystallization. Likewise Franklin, in experiments replicating the activity of building up copper artifacts by folding layers back upon themselves, reported that the process of cold hammering alone simply could not account for the metallographic cross-sections of the artifacts she examined from Arctic and subarctic Canada (Franklin 1982:50).

Smith's replication of prehistoric copper beads and the associated role of annealing is of interest because beads were produced by some of the most ancient prehistoric copper workers worldwide. Replication of their characteristics can illuminate how the sequence of metalworking knowledge was actually acquired and put into use. Smith discovered that thin strips of copper could be coiled into beads rather easily. Those coiled in a cold-worked state, however, bent at points of relatively high metallic stress called yield points. According to Smith, "annealing completely changed this behavior, and smoothly coiled curves could be obtained" (Smith 1965:239). Yet an early bead from Ali Kosh, Iraq, showed yield points, implying that they were not annealed. Later, Schroeder and Ruhl replicated some of Smith's experimental procedures and were able to produce a nonannealed bead without yield points, suggesting that although yield points indicate lack of annealing, their absence does not mean annealing was necessarily practiced. This result needs a caveat; the nonannealed, nonyielding bead produced by Schroeder and Ruhl was formed while held firm in a vice (1968:167). Their conclusion was that although not absolutely necessary, annealing was important for "easy working into a complex shape" (167). Likewise, Vernon observed that "rolled pieces may or may not reveal yield points" (1990:505).

Schroeder and Ruhl's work on artifacts from the Great Lakes region and Tennessee documented that deliberate annealing sequences were commonplace. Annealing took place at temperatures likely higher than 600 C and in some cases probably approached 800 C. Their experimental data suggested that recrystallization of the metal was accomplished completely when the copper specimens were heated to 500 C and held there for 15 minutes (1968:167). Final anneals, as determined from actual artifactual evidence, seemed to have taken place at higher temperatures than those minimally needed to induce recrystallization. Though some artifacts were left in a softer-than-needed condition by this process, others were again hammered to produce final shaping as well as to harden specific areas such as working edges. Likewise, Wayman and colleagues discovered that microstructural characteristics of artifacts they examined from Ohio Hopewell sites collected by Squier and Davis (1848) suggested lengthy final anneals at temperatures in excess of 600 C (Wayman, King, and Craddock 1992:118). It was unclear, however, whether these anneals were intentional manufacturing events or the results of crematory fires.

Clark and Purdy's 1982 work examined the microstructures produced before and after a range of treatments was applied to copper. They attempted "to evaluate the effects of open fire heat treating on the physical properties of native copper both prior to and after being cold worked" (1982:47). During initial hammering, malleability decreased rapidly and the temperature of the copper increased to a level that became uncomfortable to handle. Radial cracks appeared and spread. During the subsequent anneal they attained a maximum temperature of 732 C after burning a small fire of yellow pine for 15 minutes. Both hammered and nonhammered samples were heated to that temperature over 30 minutes' time. Ductility improved markedly, and it became clear that in order to produce sheets of thin copper several anneals would be necessary (47). Upon microscopic examination they found that copper grains had been restored "to a more equiaxed state" but that "grain size is smaller than that of the as-received native copper" (52–53). In other words the anneal did not restore the hammered copper to the state of unworked metal. They concluded that an appropriate annealing event yielded fine grain structures and was temperature controlled well enough to reduce unwanted distortions in surface texture. The copper artifacts that they examined exhibited variable hardnesses suggesting different final treatments. A bead from Florida exhibited no evidence of final annealing, but a copper sheet from the Poverty Point site in Louisiana appeared to have been subjected to an anneal after final shaping.

In the largest comparative study to date of copper artifacts (n = ca. 600), representing at least three millennia of North American prehistory, Leader (1988) documented the variable use of annealing in the copper-working practices of three different aboriginal traditions. He compared the technologies practiced by the craftspeople of the northerly Late Archaic so-called Old Copper Culture of the third to first millennium B.C., with those of the well-known Hopewell (400 B.C.–400 A.D.) culture of central riverine North America and those of the later (ca. 1000–1500 A.D.) Mississippian cultures of the American southeast. Annealing played an important role in the shaping of virtually all of the tool and ornament forms representing the Late Archaic copper-working cultures, with the exception of beads. Although final spot hardening/hammering was practiced on some early tool forms, it was not necessarily functional, if functional was defined as prolonging the use-life of the tool. Experimental data suggested that hardened tools broke and were lost to future uses, while softer ones merely bent and could be repaired (55–57). Leader's study of Hopewell site artifacts and those from Mississippian sites left no doubt that annealing processes were essential in the reduction of copper to foil thinness. These skills were so delicate and complex that Leader concluded that they could only have been accomplished by specialist artisans (198–99).

Looking exclusively at artifacts attributed to the Old Copper Culture (n = 62 specimens), Vernon (1990) alleged that it was possible to reconstruct in great detail ancient fabrication processes, largely those of hammering and annealing: "The effects of annealing are seen on many artifacts and its seems probable that ancient artisans were aware of the effects of heat on the metal during the manufacturing process, and most likely annealed their artifacts in open wood fires" (1990:502). Studies of relative hardness during annealing experiments showed that well-annealed artifacts were nearly as soft as native levels, that is, as soft as specimens that were never hammered.

Other researchers verified that annealing procedures were common throughout time and across distant parts of northern and eastern North America. They appeared to be highly variable in application from one artifact to the next, as well as within and between sites, regions, and time periods. While Childs (1994) found that Late Archaic beads from Vermont and other eastern sites were partially to completely annealed during manufacture, Leader found this not to be true of Late Archaic beads from the Great Lakes region, nor of tube beads from the Etowah site (1988:135–6). Sutter (1993) and Stevens (1996), working on copper artifacts from the Initial Woodland site 20KE20 in northern Michigan, found evidence of variation in annealing uses within the same well-provenienced collection. Beads appeared to be annealed, while bead blanks were not, which suggested that annealing occurred late in the bead manufacturing sequence. A copper projectile point from the same site showed a highly deformed grain structure with no evidence of final annealing (Sutter 1993). It is reasonable to conclude that annealing procedures were applied situationally to particular specimens of raw material, depending upon elemental composition, prior conditions of stress, and relative freedom from entrained material. Diffferent technological styles might have been responsible for differential treatments of artifacts. It is also certainly the case that the purpose to which an artifact was to be put must have dictated in part the annealing sequences to which it was subjected.

Other Means of Shaping Copper Artifacts

Grinding is a general manufacturing technique that groups several distinct operations and/or different goals: grinding for metal reduction/removal, grinding for beveling/sharpening, and grinding for uniformity of thickness. Pleger commented on the importance of grinding to the copper workers of the Late Archaic industries of the Lake Superior basin: "Grinding is one of the first techniques (along with hot and cold hammering) applied to native copper in the Great Lakes . . . virtually all finished harpoons, knives, spear points, celts, spuds, etc. exhibit evidence of grinding in the form of grind striations on the lateral edges of the implements" (Thomas C. Pleger, personal communication, 1996). Grinding away the metal was accomplished by various kinds of sandstone and other rock, probably with a range of shapes, sizes, grits, and hardnesses. Wittry (1957:205) determined by examining both pieces of waste and individual artifacts that grinding for reduction was a frequently used technique in the manufacture of the larger implements thought to typify the Old Copper Complex copper-working tradition. This finding was reinforced by Leader (1988), who identified it as the method used with hammering to rough out the shape of a tool, to shape bevels and ridges on the central and edge portions of assumed spears and projectile points, and to shape the sockets for hafting.

Edge beveling and sharpening were produced by grinding. Griffin (1961:Plate XX) depicted the variable patterning that the bevels assumed; both double bevels and asymmetrical single-edged beveling were common. Likewise beveling occurred on one or both tool faces, probably depending upon the intended functions of an implement. Grinding for sharpening occurred with great frequency on the tips of awls or piercing implements, on the working edges of celts and chisels, and in fact on the working edge(s) of virtually every tool form. "The beveled edges are thinned to a knife edge which is worn and serrated" (95). Penman, on the other hand, studied surface striations, especially at edges, as evidence of utilization scarring, wear, and reworking, presumably resulting from grinding (1977:6). He made little comment about the manufacturing status of bevels, but he did interpret some longitudinal striations as "the product of surface finishing" (12). These differed greatly in appearance from formal bevels.

About half of the projectile point varieties in Wittry's typology included ground ridges. The central ground ridge was the most commonly occurring feature attributable to grinding for reduction; Leader interpreted this feature as a strengthening or supporting "rib," an engineering solution to the inherent weakness of otherwise rather thin and soft projectiles (1988:58). Final definition of a ridge, as Leader suggested, may have resulted from grinding: "In the case of the spear points with a triangular cross-section, the central rib was formed by hammering and then refined by grinding" (53). In any event, the ridges were markedly symmetrical, which would be simpler to accomplish by grinding than by any other available means.

Grinding was also the means by which socket elements on Archaic Tradition spear points and other projectiles and implements were formed. The socket was formed by careful folding, rolling, and hammering, and then excess material was trimmed by grinding. This probably occurred in several stages, beginning at the basal corners of the implement. Presumably the so-called notched spear points described by Leader were produced in a similar way, as were socketed celts and knives (59).

In contrast to these earlier artifacts from the northern Great Lakes, Leader described grinding as the "primary technique of choice for the majority" of Middle Woodland axes that he examined (80–81). He characterized grinding as an excessively slow means of tool production and one that represented a significant departure from earlier practices. Why this change in tool manufacturing techniques occurred is an interesting question for future consideration.

It was from the annals of the earliest replicators, Cushing and Willoughby, that other aspects of the role of grinding were best documented: grinding for sheet uniformity and grinding for embossing and perforation. These techniques and their outcomes were best represented by the complex artifacts produced by Hopewellian and Mississippian craftspersons. Both Cushing and Willoughby described how careful grinding between very flat smooth rocks allowed the production of sheets of uniformly thin copper sheets. Willoughby found that "after the copper had been reduced to the required thinness, it was smoothed by being lightly beaten and rubbed with a round stone, and afterwards placed upon a flat rock where the remaining irregularities were ground away by aid of a flat stone having good grit. Fine beach sand was also employed in connection with the stone, but I do not think it facilitated the work to any great extent" (Greber and Ruhl 1989:129).

Cutting techniques were first illustrated by Cushing in 1894. This method deserves detailed explanation for it is the most likely way in which intricate cut-outs of copper were produced by Hopewell and later copper workers. Taking a smoothed sheet of copper prepared by hammering, annealing, and grinding to uniformity, the maker incised a desired figure or motif into the soft metal, leaving a raised boss on the opposite side. Leader suggested that deer antler tines were the likely embossing tools; they were shown through experiment to leave a characteristic line when pressed into copper sheets (1988:137). The boss was then ground away through gentle abrasion, leaving a characteristic square edge and freeing the design from the surrounding copper. Interior designs were cut away using the same technique. Corners and circles posed added difficulties which were solved by twisting bits of metal away, leaving a characteristic scar seen both on original artifacts and on the replicas produced experimentally (103). Willoughby mentioned the experimental use of a stone incising tool and careful bending of copper sheet to effect cutting and trimming to a desired shape. He claimed to have used a sharp stone to perforate sheet copper rather than abrasion methods: "[T]he perforations were made by using a rudely chipped flint as a drill and reamer" (Willoughby 1903:56). In contrast, Pleger and LaRonge experimentally produced rivet holes and perforations for pendants using what they termed the *punch-grind punch-drift* technique, terms borrowed from the blacksmithing industry (Michael LaRonge, personal communication, 1997). Such holes were produced in an identical way to the embossing-abrading sequence explained by Cushing, up to a point. But the last stage, that of squaring and drifting the hole to the proper form, was produced using a square awl or punch of copper (Thomas C. Pleger, personal communication, 1996).

Polishing and burnishing refer to processes that create shiny and/or smooth surfaces on copper implements and ornaments. This could be the last operation in the production sequence, but it might be more accurate to call this treatment a maintenance activity as well as a production phase. In fact Leader went further and proclaimed that shining of copper artifacts probably had ritual significance in some prehistoric Mississippian societies. To Leader, the need for polishing was associated with copper's active role in living ritual rather than solely its passive role in ritual cremation or burial (1988:140).

Polishing and burnishing as final work were practiced throughout the time span of prehistoric copper use. The most ancient use of polishing comes from the repeatedly observed careful finishing of prehistoric copper beads, whose joining or abutting seams were routinely polished to the point of near invisibility (Childs 1994; Greber and Ruhl 1989; S. Martin 1993). The likely polishing media, according to the experimental findings of Willoughby, were fine sand and wood ashes (1903:56). Leader documented fine burnishing in the joins of complex copper artifacts from Hopewell-tradition sites (1988:99). Greber and Ruhl, in their study and partial republication of C. C. Willoughby's work at the Hopewell site, reported essentially the same degree of care in metalworking: "[I]t is probable that the overlapping edges were very carefully ground and burnished until they were barely perceptible" (Greber and Ruhl 1989:100, 121). Willoughby described the sequence of such polishing and analyzed its frequency. At Hopewell, ornaments were measurably thinner at the narrow outer edges than at the wider centers, which suggested that the polishing had taken place following the cut-out stage of production. Polishing striae were sometimes visible on both sides of ornaments and were seen to follow the contours or directions of the cut-outs; in other words polishing occurred after the production of intricate shapes (153).

Sinking, molding, forming, and cladding are additional means of shaping some artifacts. The first three are synonymous terms referring to the means by which sheet copper was pushed and gently hammered into or around a precarved wooden or sandstone form and then extracted while maintaining the desired shape. Such techniques were used frequently by Hopewell people, but the general technique was likely much older. For example, the symmetry and common shapes of many of the socket designs of Late Archaic cutting implements suggested to many researchers that the sockets might have been hammered around predesigned forms (Leader 1988; Smith 1965; Steinbring 1975). Perhaps the shafts upon which the tools were to be attached were these forms. Certainly this was also the case with beads, some of which may have been wrapped or wound around a form (sometimes called a mandrel) to effect their round or quasi-round shape. The mandrel may have been something casually acquired, such as a convenient stick or bone of a desired size. Techniques of sinking, molding, and forming were commonly used, particularly for the production of ornaments and decorative emblems, from at least Middle Woodland times onward.

Squier and Davis assumed that the compound copper earspools of the Hopewell era had somehow been pressed into shape and remarked that their excavations revealed sandstone molds they believed were associated with earspool manufacture (Wayman, King, and Craddock 1992:99). C. C. Willoughby described the replication of a Hopewell-like copper earspool, using sheet copper over a wooden mold. The sheets were molded over the form by gentle hammering and pressing, using a bone to apply pressure as needed. Intermittent anneals aided this procedure. The result was a pair of earspools made of four identical copper discs. While Wayman's research group favored Willoughby's conclusions (133–34), Ruhl debated the reality of this method. Based upon the variation in size and apparent construction techniques she observed while measuring about 500 Hopewell earspools, Ruhl concluded that the earspools were formed freehand: "There is no question that Willoughby demonstrates that the use of wooden molds is technically possible, but the evidence of the artifacts is otherwise" (Greber and Ruhl 1989:141).

Hollow, domed bracelets and brooches of Hopewell times were likely molded or sunk in molds consisting of shallow depressions in stone or wood. They assumed the form of the mold by gentle hammering on the interior of the ornament. Leader suggested that the pronounced symmetry of the bracelets in the Hopewell collection were perhaps the products of a single workshop or maker (1988:86). Copper dishes of Mississippian times were likely made in the same fashion.

Cladding was a related process, in which a precarved form of a different material was covered or laminated with a layer of thin copper sheet. The cladding of wooden, clay, or shell materials with a cover of copper or other metals was a feature of Hopewell technology but became more common in Mississippian times. Studies of Hopewell-era artifacts upon which cladding was practiced suggested that the metal caps were pressed and burnished tightly and perfectly to fit the underlying forms. Leader (137–38) suggested that, at least during Mississippian times, sharks' teeth were used to carve the wooden forms upon which cladding appeared.

The cladding processes were elaborate. Sometimes separate copper sheets were folded and overlapped to attach them to one another as well as to secure the wooden half-forms over which they formed cladding. This work was skillfully demonstrated by some of the compound wooden rattles clad of copper from the Etowah site (Leader 1988:131). Artifact forms that appeared earlier in Hopewell times, such as the earspools mentioned earlier, were also produced in Mississippian times, but they were often objects of other materials clad in foil rather than made exclusively of copper sheet. This change suggested that efforts were made to conserve the use of copper. Alternatively, perhaps the exploitation of new sources of copper with fewer impurities allowed the production of ultra-thin foil (110).

What happened to individual fragments of copper when they were too tiny to be turned into usable objects? The answer, for cultures that practiced alloying, smelting, and casting of metals, was simple. They added them together and combined them by raising the temperature to their melting point(s). The prehistoric copper workers of eastern North America did not use smelting or casting techniques, so making use of small scraps of copper was done in other ways. Part of our fascination with the question of scrap utilization probably has to do with our perceptions of copper as a scarce commodity. But it is clear from microscopic studies of copper that this concern, for whatever reason, was obviated by the methods of the prehistoric users.

Most commonly, the small, platey extensions that protruded from a piece of copper as it was shaped with a hammer were simply folded under and reincorporated into the mass before they broke away. This practice leaves a characteristic layered appearance, which is sometimes visible without any magnification. A closer look shows that even the smallest beads were composed of what appear to be layers (fig. 22), as these surfaces were folded

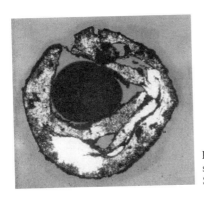

Fig. 22. Copper bead cross-section at 25x, showing layering of copper. (Photo by John Stevens.)

back into the object (Childs 1994; Stevens 1996; Sutter 1993). But such layering also occurred when multiple pieces of copper sheet, both large and small, were combined by hammering them into a solid mass. Sometimes even objects as small as beads were composed of more than one piece of copper. Childs (1994) illustrated one such bead, in which the end of one fragment was secured within the grip of a double-sided piece of another fragment. Some fusion through the careful control of heat had occurred to ensure a degree of adherence between the separate pieces of copper.

In the North American subarctic and Arctic regions, many tools were routinely built up of separate fragments of copper by folding and rolling smaller sheets together (Franklin 1982). Subsequent hammering consolidated the pieces into a bar or other shape. Leader documented the folding and consolidating technique in the Upper Great Lakes region and suggested that its use was not typical of Late Archaic cultures but those of more recent groups. Such conclusions might be taken more literally were there well-dated artifacts to provide the required evidence of technological differences through time. Leader's conclusions about the relative antiquity of the methods of rolling and consolidating may be considered only provisionally correct for the Upper Great Lakes region. The method was certainly in use by Late Woodland times (McPherron 1967:167).

Riveting (fastening a piece of copper to another material or to another piece of copper by means of a copper pin or roll) was practiced in earliest times in the Upper Great Lakes. Large copper tools were attached to wooden or bone handles by hammering copper rivets through the tool and into the handle. Other forms of riveting were in practice during Hopewell times, and were common later on, by which time people began to fabricate three-dimensional hollow objects from sheet copper. In some cases, such as seen in some Hopewell earspools, two or more sheets of copper were joined by means of a central rivet of rolled sheet. In some cases the exterior sheet edges were turned under and crimped around both lips of a funnel-shaped central rivet. In other earspools, the rivet was inserted and flanged through central perforations in the inner sheets, and the exterior sheets were lapped over the inner sheet, covering the rivet flanges. On Mississippian artifacts from the Etowah site, rolled rivets of equal size were inserted through perforations in two pieces of copper sheet, then flanged to keep them together. Sometimes it appeared that this was done to repair broken sheets; other times it was done to mount a strengthening sheet to the back side of a decorated sheet to create a socket for a support (Leader 1988:116). Other artifacts were riveted together to repair damage to the copper sheet, which suffered metal fatigue from repeated bending/flexing and perhaps excessive polishing.

Some very conjectural methods of working copper were suggested by other experimenters, but the degree to which they were actually practiced by prehistoric copper workers remained undemonstrated by artifactual evidence. Hoy's 1908 paper described two experimental methods, rolling and swedging, which he claimed were used to produce ancient copper tools. Rolling (discussed above) was the process by which a malleable lump of copper was placed between two stones, and in the manner of a modern rolling mill was compressed into a cylindrical shape. Swedging was allegedly accomplished by driving material into a form or mold and effecting a change in its shape. Hoy wrote, "Some of these implements that have been supposed to be cast were swedged; that is a matrix was excavated in stone, into which the rudely fashioned copper was placed, and then by repeated blows the article was made to assume the exact shape of the mould" (1908:172). Hoy believed that swedging accounted for the plano-convex (flat on one side and curved on the other) shape of copper artifacts commonly found in Wisconsin surface collections of his era. He suggested that the ancient craftspeople sometimes used a double mold to shape more complex objects. In this method, there were two stone swedge forms, and the metal was placed between them and hammered until it acquired the desired shape. He claimed to have found such a half-mold excavated in a large granite boulder. Steinbring used the same swedging terminology as Hoy (Steinbring 1975:95) and claimed that swedging was the likely means for the production of Middle and Late Archaic spears that displayed central ridges. But his actual description of the technique was the rolling process Hoy first described. Examinations of large collections of Middle and Late Archaic artifacts do not support the use of swedging as a manufacturing technique (Leader 1988; Penman 1977; Vernon 1990; Wittry 1957).

Dividing is a cutting technique Steinbring suggested may have been used to produce "twin" copper tools. In dividing, a large sheet was prepared and a rough shape that was bilaterally symmetrical was produced on it by hammering, yielding a shape of "a long flat blade with gradually tapering points at both ends" (1975:95). The center was gradually dented with a heavy stone baton and eventually was severed at the middle by heavy blows from an adze. This resulted in two identical spears with partially prepared haft elements. Although this method is conceivable, the final trimming of the haft and the edges would obscure any evidence of the dividing process. Material evidence for its use is lacking in the artifacts reported in the literature.

The last technique, drawing, refers to stretching thin wire-like strands of ductile copper. The method was described and attempted by Cushing in 1894; he learned it from Zuni metalworkers of the southwestern United States. Recall that Cushing showed that the production technologies of non-industrial metalworking were adequate to produce all of the characteristics displayed by the metal artifacts of eastern United States prehistory. The likelihood that this method was routinely used in the east is of great interest, because it could account for the fabrication of very tiny copper beads whose production technology is not well understood (S. Martin 1994).

To draw copper, the worker began with a lump of raw copper, which was hammered into a long, thin rod. The rod was then twisted, pulled, and pushed through a series of graduated perforations in a drawplate. The drawplate was generally made of the scapula or horn of an animal; the drawing operation was aided by the liberal application of grease or fat to the drawplate perforations. Cushing claimed that by using this method he was able to produce "copper wire as fine as coarse linen thread" (1894:99). According to Cushing, the technique itself, derived from earlier general methods of working other materials, particularly shaft straightening. It would be of interest to replicate copper wire by this method and to conduct periodic metallographic inspections to catalog characteristics of grain deformation during the process.

Were Prehistoric Artifacts Worked at High Temperatures?

The question of how hot the prehistoric copper workers were able to heat their metals during its fabrication into useful forms has been asked repeatedly (Childs 1994; Clark and Purdy 1982; Frank 1951; McPherron 1967; Schroeder and Ruhl 1968; Steinbring 1975; Vernon 1990; Wayman, King, and Craddock 1992). This is a fundamental question, which really asks whether there was a North American genesis of true metallurgy that aimed to change the chemical composition of materials as well as their physical characteristics. Were the metals hot forged? Was copper heated to a white-hot or hot-short (just under melting) temperature? Were forced drafts of air used to provide critical volumes of oxygen? Were layers or fragments of metal accidentally or purposely fused together? Were multiple metals accidentally or purposely mixed? Were the artifacts from so-called crematory altars of Hopewell times intentionally or accidentally melted? It seems reasonable to address each of these lingering questions with a short summary of the available evidence.

The hot forging question can be quickly dispensed with—"Probably yes, but indeterminate!" It seems that metallurgical analyses cannot easily distinguish between the characteristic grain deformations of cold-hammered/annealed versus hot-hammered metal (Clark and Purdy 1982; Frank 1951; Franklin 1982; Schroeder and Ruhl 1968). As Vernon put it, "Hot working is an alternative to cold working and annealing, with similar end results in microstructural features" (1990:502). Clark and Purdy concurred: "Microstructural distinction between these two processes is nearly impossible, and it is not presently known if hot working was used in the fabrication of any of the North American copper artifacts" (1982:47). They were able to demonstrate that a small wood fire quickly raised the temperature of a copper nugget to greater than 700 C.

Overall, opinions actually ranged from skeptical to enthusiastic regarding hot working. Vernon doubted its efficacy in Archaic times; hot pieces were hard to hold onto, no oxide inclusions were reported, and directionality of nonmetallics was also not reported (1990:502). Childs suggested, however, that quick hot forging may have been responsible for the microstructures in a bead that she examined from the Boucher site in Vermont (Childs 1994:239). So we are left with indeterminacy to date. A systematic look at a range of artifact types to search for Vernon's distinguishing characteristics (i.e., presence of oxides and directionality of nonmetallics) might put this question to rest. Examination of the archaeological record for evidence of objects or material that could serve as tongs or holders for hot metals might also be of interest. Spring-like tongs are relatively simple to fabricate from wood. We really do not know much about the range of perishable tools that people of ancient times were able to make and use, and this is one instance in which we may have to confess our patent lack of knowledge about the full range of prehistoric technological alternatives.

Welding or joining fragments of metal together, assisted by hot enough temperatures to effect a permanent attachment, was not routinely thought to be an eastern United States prehistoric technology. However, several researchers looked carefully at the evidence for such fabrication methods. McPherron began, in 1967, to investigate such possibilities. At the Juntunen site, in Mackinac County, Michigan, it was clear from three seasons of excavation that copper-working technologies were well represented in the materials discarded at the site. In fact it was the most well-endowed Late Woodland site in the region as far as native copper was concerned, and the many small fragments recovered there inspired the idea that they were amalgamated somehow to make them useful. Subsequent experimentation with well-cleaned industrial copper failed to achieve the welding of copper fragments. "It does appear that oxides will prevent at least the kind of copper we used from joining (i.e. in the way that iron can be welded by forging at white heat), at the temperature available without bellows or other specialized equipment" (McPherron 1967:167–68).

Obviously the failure of twentieth-century archaeologists to imitate the assumed methods of twelfth-century artisans does not put the question of welding to rest. Better evidence for its importance must come from the ancient artifacts themselves. Childs (1994) reported on such work, this time investigating the fabrication characteristics of copper beads from the first millennium B.C. Boucher Cemetery in Vermont. In one bead, a magnified cross-section view clearly showed that the bead was actually composed of more than one piece of copper. Childs suggested that this was likely done at high temperatures to "maximize the chances of metal fusion or pressure welding between layers" (240–41). What is even more interesting is the fact that in two other beads from east coast localities, Childs identified probable pressure weld lines, or lines of oxides that would indicate that the bead had been worked at high temperatures. She also discovered microscopic evidence of cracks around which some fusing of metal had taken place (242). The story of high-temperature copper working is not yet finished and will be more fully told when additional comparative work is done on beads and other frequently encountered copper artifact types. Beads are particularly suited for such comparative work, for they occur over a wide area and through long reaches of time, providing a potentially large data set from which to examine evidence for changing technologies and technological variability. Such studies might show that innovative technologies replaced traditional ones in some places, or that a range of technologies existed side by side, to be used as conditions required or suggested them.

The most extensive study of high-temperature treatments of prehistoric metals was begun by Wayman and colleagues, who conducted metallographic examinations of ca. 121 artifacts of Hopewell-era copper and other metals from the collections of Squier and Davis (Wayman, King, and Craddock 1992). This research addressed a complicated set of questions, including examining claims that some of the Hopewell artifacts were cast of bronze. There also were fundamental problems of provenance to be resolved before metallography could commence with any substantial results. Materials from a range of sites, including some in South America, were mixed with the Ohio materials. Furthermore, many of the artifacts studied by Wayman's group suffered postdepositional heat damage, probably the result of intentionally set crematory fires. Thus their utility for resolving questions of high-temperature manufacture (welding, bonding, brazing) was somewhat limited, because the important evidence in the form of original microstructures was somewhat compromised. Wayman and colleagues commented that in fact very few studies of microstructural characteristics of prehistoric native copper metalworking had been done and the true complexities of regional technologies were not completely understood. But even with these caveats, Wayman and colleagues' results bear attention.

Several of the Hopewell culture copper artifacts were unusual in elemental composition as well as in physical characteristics. In at least one case, a natural alloy of arsenic and copper (algodonite) had been worked into the form of an earspool. This brittle material was harder to work than other more pure nuggets of copper, but had the advantage of a distinctive color, which Wayman and colleagues implied was intentionally sought after.[2] The most interesting artifact from the perspective of studying high-temperature copper working was a dome-shaped button or bead in which two distinct layers of copper apparently enclosed a third arsenic-rich layer of copper. The internal layer appeared to be different enough in elemental composition to render a lower melting point, thus its utility as a bonding material: "It is clear then that there are significant differences in the metallurgy of the bond region as compared to what appears to be unmelted native copper sheet joined by the bond" (129). Wayman and colleagues' dilemma was that although this bonding could not be shown to be intentional, the conditions to create it were too precise to be accidental. While granting that the ancient people must have known about the variable properties of such natural alloys as arsenic-rich algodonite, Wayman's group concluded, as reasonably skeptical investigators should, that they could not account for the metallographic appearance evidenced by the quasi-bonded artifact (133). The prospects for learning more about the intricacies of high-temperature artifact fabrication are great; they depend on painstaking microscopic examination of large comparative data bodies. Researchers are beginning to define what it is they do not know about the complexities of prehistoric North American metalworking.

In many parts of the world, prehistoric people invented metal smelting (heating to molten temperatures enabling the extraction and mixing of metals, removal of impurities, and production of alloys with new physical properties). If prehistoric cultures discovered the ability to control a heat source for melting native copper, the rest of the technology of smelting probably existed in a nascent state as well. Researchers suggested that virtually everywhere in the world where such melting happened, smelting followed quickly. The melting/smelting technology may have been accidentally discovered in pottery-making cultures as a side effect of firing painted or glazed pottery vessels in kilns with managed oxygen flow (Tylecote 1992). Metallurgy does not appear in all such cultures, but so far as we know it rarely occurred without parallel technologies for clay. In prehistoric North America such technologies were practiced in areas of present-day Mexico and near the southern borderlands of the United States, but neither technology arose, so far is as known, in the northeast. Tylecote (1992) and Craddock (1995) provided recent accounts of comparative histories of metallurgy.

In the case of native copper, temperatures in excess of 1083 C must be reached to achieve melting. This temperature is marginally beyond the upper range of a wood fire; to reach it one must fire the copper with fuel other than wood, such as charcoal (Craddock 1995:122). Alternatively, one must provide an artificial draft to encourage adequate combustion.[3] Tylecote provided data on what sort of drafts might have been used in early smelting technologies. "It has been generally accepted that the human being is capable of blowing intermittently at a rate of 40 L/min. which should be enough for two blowers to maintain a well-insulated furnace at a temperature of 1200 C. Three blowers could probably have maintained a continuous supply of air" (1992:15). Depictions of preindustrial blowers sitting at small smelting furnaces delivering air blasts via pipes are commonplace in metallurgical cultures, where they occur in prehistoric art, in sculpture, in historic documents, as well as in archaeological and ethnographic records.

Archaeological evidence for melting, smelting, and casting of metals in prehistoric eastern North America has not been found to date. Metallurgists began the search in 1848 for such evidence, when Squier and Davis requested that a Professor Church perform a quantitative analysis on fused masses of metal found at Mound City in Ohio (Squier and Davis 1848). Church determined that the metal was not bronze, but copper. But despite Church's discoveries, the false allegations that portions of Squier and Davis's finds were bronze occasionally found new life among writers intent upon connecting North American native copper with the European Bronze Age (Fell 1976:129).

It has already been stated that the Squier and Davis collection, which was the historical product of the collecting activities of at least one British citizen and sequential ownership by several museums (ultimately the British Museum), suffered from problems of provenance. Some artifacts in that collection were purchased in Peru, and others were purchased from Squier and Davis in North America. Confusion began when Davis identified one of the artifacts from his collection as bronze. The provenance of this article was not immediately clear. However, the catalog treatment of the specimen was actually consistent with the treatment of other artifacts from various parts of South America. The most probable source of the confusion was that no one kept very accurate records about where and from whom artifacts were acquired.

Wayman and colleagues' analyses helped clarify some of these problems by identifying other artifacts, never mentioned by Squier and Davis, which probably were added to the collection over its 150-year history of ownership. Here is the research group's assessment of the particular bronze artifacts in question, which were referenced by acquisition unit and catalog number: "The presence of bronze in a Hopewell assemblage would be of extreme interest. However, unfortunately it is far from impossible that all of these S556 artifacts could be Peruvian, possibly being represented by CAT items U22 and U25. In any event, the association of these S556 artifacts with the mound copper must be considered highly questionable" (Wayman, King, and Craddock 1992:117).

To Wayman's group, problems of provenance, inconsistencies in technological approach, inconsistencies in style, and the failure of Squier and Davis to mention the artifacts in lists of materials taken from Ohio all pointed to the bronze artifacts' deriving not from Ohio but elsewhere. By 1992 they were somehow stuck in the same collection.[4] Wayman and colleagues' conclusion, based on comparative assessment of all 121 artifacts in the collection was straightforward: "Thus there is no evidence whatsoever for smelted metal in the Hopewell material" (133). Notice that they did not refuse to look for the evidence. They looked, hopeful that such evidence would be extremely interesting, but were disappointed.

Listing all of the scientists who have looked for and failed to find such evidence is important, because such an exercise dispels the notion that somehow archaeological science does not wish to question standing models of prehistoric copper working and acquisition. The facts of the matter are otherwise. From 1951 researchers have quietly and slowly performed the microscopic analyses that are necessary to demonstrate the presence/absence of such processes as melting, smelting, and casting (Childs 1994; Clark and Purdy 1982; Drier 1961; Frank 1951; Root 1961; Schroeder and Ruhl 1968). To date their results are remarkably and notably consistent. *No one has found any evidence that points to the use of melting, smelting, and casting in prehistoric eastern North American copper working.* Nor has anyone excavated prehistoric features identified as kilns or furnaces, nor have any prehistoric slags been reported. While recognizing that their knowledge is incomplete, researchers are unanimous and comfortable with the validity of their provisional conclusions: no evidence for intentional melting, smelting, and casting.

Conclusions and Further Questions

The ability to understand the processes by which prehistoric copper artifacts were produced is hampered by the relatively small body of data from which conclusions are drawn. Only ca. 1100 artifacts have been examined carefully to explain the means of their manufacture; many of these were examined visually rather than metallographically. Small samples from limited areas of northeast North America hamper our complete understanding of the range of internal structures of prehistoric metal artifacts. Granted, metallographic cross-sections are expensive in terms of preparation and analytical time and require expensive instrumentation and special training to accomplish. They are also invasive; that is, sometimes delicate and rare artifacts must be damaged in order to collect a study sample of the metal. Additional metallographic analyses of artifacts from many different times and places are needed to understand completely the production methods that prehistoric people discovered and practiced on copper and other metals. In addition, more attention should be paid to the externally visible characteristics of large collections of copper artifacts, particularly their measurement, a need pointed out by a number of other researchers (Childs 1994; Leader 1988). In statistical modeling, it is a truism that the smaller the sample in the face of high levels of variability, the less secure one can be that the characteristics seen in the sample actually reflect those of the whole population. This is true of our knowledge of copper; we need to look further and at more materials to have the fullest confidence in our conclusions.

The inability to distinguish between cold-hammered/annealed versus hot-forged copper is a perplexing situation. Fortunately, experimental replication can help fill the void. Despite the fact that replication can never provide more than an inferential conclusion about the actual techniques of prehistoric copper working, it may provide a broader set of hypotheses or probabilities than metallography alone. Systematic investigations of changes in copper subjected to experimental hot working may be a valid source of new inferences about the prehistoric feasibility of such techniques and their changing application over time. Craddock remarked that experimental results succeeded in enlightening present-day researchers about little-known or poorly recorded preindustrial technologies (1995:22). More experimental replication will benefit our understandings, especially in conjunction with larger samples for macro- and microscopic studies. In addition, the data for documenting prehistoric copper working come not only from artifacts but from the production sites where the processing of metals took place. Careful professional investigation of such places is essential before the story of prehistoric copper working can be completely told.

Suggestions for Further Reading

Cushing, F. H. "Primitive Copper Working: An Experimental Study." *American Anthropologist* 7 (1894): 93–117. Classic account of Cushing's experimental methods in replicating native copper artifacts of Hopewell affiliation.

Greber, N., and K. Ruhl. *The Hopewell Site: A Contemporary Analysis Based on the Work of Charles C. Willoughby.* Boulder: Westview Press, 1989. An archaeologist and a metals analyst assess Willoughby's original fieldwork and metalworking experimentation.

Tylecote, R. F. *A History of Metallurgy.* 2d ed. London: Institute of Materials, 1992. Worldwide survey of the evolution of metallurgical techniques with great detail on earliest smelting evidence.

Vernon, William W. "New Archaeometallurgical Perspectives on the Old Copper Industry of North America." Geological Society of America *Centennial Special Volume* 4 (1990): 499–512. Comprehensive account of copper-working techniques of the Archaic Tradition in Wisconsin and Michigan.

Wayman, M. L., J. C. H. King, and P. T. Craddock. *Aspects of Early North American Metallurgy.* Occasional Paper 79. London: British Museum, 1992. Three studies of native North American metalworking from Ohio, the northwest coast, and the Canadian Arctic.

6

They Must Have Been Numerous, Industrious and Persevering and Have Occupied the Country a Long Time

—Charles C. Whittlesey, 1863

Daily Life

> The importance of a cache lies chiefly in the fact that it enables a number of different types of artifacts to be classed together in space and time. In other words, it provides direct evidence of a cultural cohesion which would necessarily remain presumptive if only collections of isolated artifacts were available.
> —Robert Popham and J. Norman Emerson, 1954

Introduction

In order to go beyond descriptions of things to more interesting discoveries, one must seek to understand ancient cultures in their entirety. One must concentrate on the reasons why artifacts were made and used, as well as the means by which they became part of the archaeological record. Some people might suggest that this is a case of spinning straw into gold, for archaeologists must continually grapple with what might be termed unacceptable levels of uncertainty. What, in fact, does one find when one digs up an archaeological deposit in the Upper Great Lakes? The sites are small and fragile compared to the cities of the classical civilizations. Many of the sites in the Upper Great Lakes are less than 400 m sq in size, particularly those that were occupied only once. Those used repeatedly might appear to be a bit bigger, maybe even on the order of half a hectare to a hectare in size. Sometimes permanent settlements or ceremonial sites, at least in Michigan and Wisconsin, run much larger; the sites of Aztalan in southern Wisconsin and the Norton Mounds in southern Michigan are two such places. Some sites in very rich reliable food-getting environments appear fairly large and complex as a result of repeated occupation.

Sites consist of superimposed deposits of soils, sands, rocks, and garbage rather than solely of artifacts or structures. One sometimes ends up moving tons of dirt and sand to uncover artifacts. Here and there lie the remains of an ancient picnic, work activity, favorite pot, toy, or baby tooth. Someone's dogs or kids probably ran through the debris as it was lying there, and as a result the food remains and broken flints or sherds can be trampled, scattered, and then perhaps redeposited again. By the time the archaeologists get around to investigating the site, some 1500 years or more after the fact, much of what was dropped has disappeared. Only a few really durable things are still recoverable as three-dimensional objects, and they may be poor reflections of a once-whole way of life. The less sturdy things are marked only by stains in the soil or by microscopic chemical traces rather than visible substance. Where a birch bark structure may have stood, a faint line in the soil may follow the edge of the ancient bark wall. A few broken artifacts may lie close to the boundary between outside and inside, helping define the edge of home to someone from 300 A.D., perhaps swept out of the way as someone lay down to rest, or to sit and talk.

This chapter discusses some of the archaeological sites of the Lake Superior basin and Upper Great Lakes in an attempt to tell the story of ten thousand years of prehistoric life in the region. It features the life cycle of copper artifacts and attempts to give them a home, that being the culture or cultures in which they were acquired, fabricated, used and/or dispersed, and eventually discarded or deposited. The life histories of the artifacts from excavated contexts help build a parallel collective history of the people who produced them. The story is incomplete and is based on vastly imperfect data. To as large a degree as possible, the story is composed from archaeologically derived in situ evidence. Only such data can provide the shapes and traces that together make up the archaeological record.

The archaeological record is filled with materials other than copper. No society exists in order to manipulate a solitary material, and the prehistoric people who used copper were no exception to this generality. Looking at the whole of their material, social, and ideological lives is necessary to understand, to the extent that it is possible, their overall approaches to life. This chapter attempts to situate the material remains within the prehistoric cultural sequence of the Upper Great Lakes. It is primarily about the prehistoric cultures that made their homes in the Upper Great Lakes and their ways of life in those demanding environments.

It is necessary to restrict the story to the cultures that probably occupied the geographical region immediate to the Upper Great Lakes (Huron, Michigan, and, in particular, Superior). In reality prehistoric research does not actually obey such artificial boundaries; nor did the user cultures of prehistoric copper, whose contacts and adventures likely took them away from the region at some times during the long span of prehistory. But for the sake of practical necessity, such a geographical boundary must be imposed.

Daily Life

The Upper Great Lakes is a logical starting point to tell the story of prehistoric copper in eastern North America. The region holds a lengthy record of archaeological research and discovery. Its prehistoric cultures incorporate a tradition of copper use nearly eight thousand years long—as long a history of metalworking as virtually anywhere in the world. The earliest known copper working on the continent is found in this region. Confining the discussion to areas within the Lake Superior drainage basin reduces environmental variation to a minimum and at least holds that dimension in relative control. This discussion focuses on the land areas that overlie the Canadian Shield (whose watercourses drain into Lake Superior) and are covered with deciduous-conifer forest transitional to boreal conditions of the sub-Arctic. The local prehistoric in situ deposits are the best, in fact the only, archive available for studying the region's earliest people and lifeways.

Getting to know prehistoric people from the archaeological remains of their artifacts is complicated. As they work, archaeologists design and manipulate what are known as *archaeological cultures* to facilitate orderly description and comparison of unfamiliar objects. Archaeologists build descriptive taxonomic units through systematic accounting of similarities and differences in the material remains they study and group them together in an approximation, or hypothetical arrangement, of cultural traditions that may have actually existed. These provisional taxa are created for convenience. They allow the practitioners to communicate with one another using a practical shorthand of comparative units. Archaeological cultures are not the same as living, directly observable communities. In fact these constructions or abstractions never existed in the way real communities do, but they still have descriptive utility. Archaeologists make use of archaeological cultures as descriptive devices at very generous levels of similarity and difference to organize and compare the products of human communities across space and through time.

Sometimes the composition and description of archaeological cultures is so consuming for archaeologists that it occupies all their time, drives their research efforts, and becomes the research product itself. This is reasonable if one's goal is to build a taxonomy and if there are valid reasons for such activities—order from chaos being a primary one. However, the boundaries of archaeological cultures can be arbitrary, and the very archaeologists who created them do not always agree on where the boundaries lie. Because looking at ancient human behaviors is the overall goal here, there is no intentional focus on refining analytical units that archaeology has designed. They will occasionally be used to guide the narrative.

Table 4 shows the general time coordinates and regional locations of some of the archaeological cultures and other taxa that have been defined for the Upper Great Lakes region. These categories are in general use by

Table 4. Chronology of culture-historical periods in the Lake Superior basin and nearby (after Anderton 1993; Green, Stoltman, and Kehoe 1986).

Tradition and Stage/Substage	Approximate Years Before Present (BP)	Ethnic Groups, Archaeological Cultures, Phases
Historic/ Protohistoric	500	Ojibwa Ottawa Huron many others
Terminal Woodland	1000	Ontario Iroquois Juntunen Oneota Lakes Blackduck Selkirk Mackinac others
Initial Woodland WOODLAND TRADITION	2000	Laurel
Late Archaic	3000	Red Ochre Glacial Kame Burnt Rollways
	4000	
Middle Archaic	5000	Old Copper Shield Archaic Squirrel River
	6000	
Early Archaic	7000	Minocqua? Flambeau?
ARCHAIC TRADITION	8000	
Late Paleoindian	9000	Minocqua? Flambeau? Interlakes Composite
Early Paleoindian PALEOINDIAN TRADITION	9500	?

several generations of archaeologists across the Midwest and offer a very rough outline of temporal and spatial differences in cultural adaptations in the region. While Paleoindian, Archaic, Woodland, and Historic refer primarily to cultural traditions of lengthy duration (on the order of 1,000 years or more), the other units, such as "Blackduck," "Old Copper," "Juntunen," and the like refer to spatially localized finer differences in life way and adaptation. Archaeological cultures are often named after distinctive artifact styles in pottery, distinctive tool types, or localities at which distinctive finds were first excavated and named.

Daily Life

Changing ecological conditions that varied over short and long scales of time tended to present challenges to human communities, which were met by changes in the communities themselves. These changes were material, social, and ideological. The fact that there was cultural variation in time and across variable environments is taken as fact by most archaeologists. It is also expected that variations on a general adaptive theme co-existed, resulting in neighboring groups of people doing things in distinctive ways. This was most certainly the case in the time periods following European contact. There is every reason to expect that differentiation and interaction of social units occurred throughout the cultural history of the Upper Great Lakes region.

How Old Is It?

How old is the oldest worked copper in the Upper Great Lakes region and where is it from? Table 5 depicts some of the earliest known radiocarbon dates from the Lake Superior basin region in association with worked copper.[1] Of the sites depicted, South Fowl Lake, Minnesota, provides irrefutable evidence of the antiquity of native copper working. At this site, investigators recovered a copper spear point with a portion of its wooden haft element still intact (Beukens et al. 1992). Under ordinary conditions, organic materials such as wood decompose fairly quickly within archaeological sites. In this case, however, the wood was impregnated with oxidizing copper. Such conditions create favorable micro-environments for wood preservation because the oxidized material is bactericidal and kills off degrading organisms before they can decompose organics. Wherever copper oxides are found, there is a good chance that organic materials are preserved close by. The date itself was the result of the accelerator mass spectrometry (AMS) approach to radiocarbon dating, and is ca. 6800 years before present (894). The hafted wood associated with the point at South Fowl Lake was a living tree or sapling about 6800 years ago, and such a date confirms that people in the Lake Superior basin were fabricating composite tools out of copper and wood nearly seven thousand years ago. This is the earliest date so far from indisputable archaeological context and association with human activities for any metalworking in North America.

The early date from the Keweenaw Peninsula, Michigan, at site 20KE20 deserves comment. At this site a fragment of hammered copper was recovered from a hearth-like feature. A small charcoal sample taken from within the feature ca. 10 cm away from the copper yielded a conventional radiocarbon date of ca. 7870 +/− 350 BP (uncal) (S. Martin 1993:175). This result implied that quickly after occupying the postglacial Lake Superior

Table 5. Early radiocarbon dates in the Lake Superior basin in association with worked copper.

Site/Date	Material	Age (BP)	(*)	(**)
20KE20 (MI) Test Unit 12 (S. Martin 1993)	charcoal	7870 (uncal)	+/– 350	+/– 700
Oconto (WI) C-837 & C-839 combined (Libby 1955)	charred wood	7510 (uncal)	+/– 600	+/– 1200
SFowl Lake(MN) Anderson #1 (Beukens et al. 1992)	wood splinter	6800 (cal)	+60, –70	+220, –180
Oconto (WI) #1 (AMS) (Pleger 1996)	string/cord	ca. 6000 (uncal)	n/a	n/a
#2 (AMS) (Pleger 1996)	charcoal	ca. 5300 (uncal)	n/a	n/a
Renshaw (ON) #3 (Beukens et al. 1992)	twined cord	5320 (cal)	+10, –25	+155, –60
Renshaw (ON) #1 (Beukens et al. 1992)	plant fiber	5310 (cal)	+20, –20	+30, –60
Renshaw (ON) #4 (Beukens et al. 1992)	wood fragments	5020 (cal)	+40, –45	+275, –150

(*) and (**) report the following figures: for cal dates they are cal ranges, and for uncal dates they are 1 and 2 sigma values. Oconto dates are from Thomas C. Pleger's ongoing dissertation research and are unpublished.

basin, native peoples discovered the special properties of copper and began to use it. Such a thesis is entirely within the expectations of Lake Superior basin researchers and appears to be borne out by a number of lines of additional circumstantial evidence. Lithic procurement patterns, ancient settlement locations, earliest copper utilization, and postglacial land form stabilization all point to the early occupation and use of the western Lake Superior basin and to the existence of the Interlakes Composite as defined by Ross (1995:244).

The more dating done on objects from sound archaeological contexts, the more compelling the evidence for early utilization of copper in the Upper

Great Lakes. It is no longer extreme to accept the use of copper in the 5000 BP range, or even earlier. There are a number of additional localities from which radiocarbon dates in the 5000 BP range have come, including a date from Burial 15A at the Hind site, southern Ontario of 4570 +/– 120 BP (Donaldson and Wortner 1995). The contested dates from the Oconto site, Wisconsin, at 5600 +/– 600 BP (uncal) and 7510 +/– 600 BP (uncal) (Mason and Mason 1961; Stoltman 1986:224) deserve reconsideration given the South Fowl Lake date (Beukens et al. 1992). Recent work by Pleger on a reanalysis of Oconto materials included the AMS dating of Oconto copper tool contexts, with two resultant dates in the 5300–6000 BP range, confirmed Beukens et al.'s suspicions that the Oconto materials were genuinely as old as Libby's initial analysis revealed. These early dates are supported by studies of the Oconto site's relict shoreline geochronology (Thomas C. Pleger, personal communication, 1996). It is possible that copper working is associated with sites of Paleoindian and/or Early Archaic technologies in the western Upper Peninsula of Michigan, in Wisconsin, and on the south and northwest shores of the Superior basin (Steinbring 1991).

It is somewhat frustrating to note that many organic materials surely survive in association with copper implements of likely great antiquity. Through efforts on many local levels, especially face-to-face education for collectors, such data can be collected and reliably analyzed to yield, finally, some tight chronological information regarding individual copper implement types and their likely age ranges. AMS dating and the fortuitous preservation of associated organics at or within copper artifacts may allow the restitution of many casually collected artifacts to their original temporal contexts. It remains to collect these data. Penman (1977) reported that large numbers (n = ca. 75) of the copper artifacts of the Hamilton collection of the State Historical Society of Wisconsin, one of the largest collections of native copper in the world, contained wood residues within haft elements.[2] With this chronological information we would be in a position to trace the technological changes and processing alternatives that probably took place in early prehistoric copper working.

The Role of Copper in Paleoindian Life

The Late Paleoindian cultural stage of the Paleoindian Tradition represents a transition in human adaptive patterns caused in part by waning glaciation and its final retreat northward (fig. 23). Quasi-modern vegetation and climatic patterns quickly began to replace the periglacial environments in areas newly free of ice cover and permafrost. Within the Lake Superior basin this transition occurred about 9500 BP, when broad areas of ice-free lands became

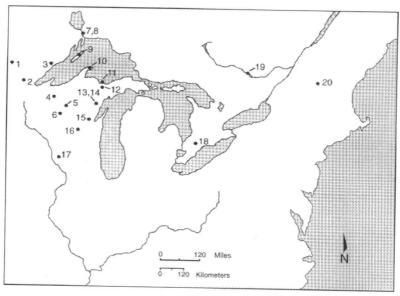

Fig. 23. Paleoindian and Archaic Tradition sites and localities: 1—Itasca, 2—Interlakes Composite (region), 3—South Fowl Lake, 4—Alligator Eye, 5—Ottawa North, 6—North Lakes sites (region), 7—Renshaw, 8—Cummins, 9—Isle Royale (Minong, Lookout), 10—20KE20, 11—Deer Lake sites (region), 12—Gribben Forest, 13—Riverside, 14—Chautauqua Grounds, 15—Oconto, 16—Reigh, 17—Osceola, 18—Hind, 19—Morrison's Island, 20—Boucher. (Drawing by Brett A. Huntzinger.)

useful to people. Pleistocene conditions lingered here longer than elsewhere; there is good evidence that the Upper Great Lakes was the last geographical stronghold of glacial ice in the northeast. Short-lived readvances separated by periods of stasis marked the glacial retreat. The Lake Superior basin was completely deglaciated by about 9500 BP (Drexler, Farrand, and Hughes 1983), from which time humans first entered the area.

A closing date for the Late Paleoindian Stage is difficult to establish, because it was a gradual rather than an abrupt transition. Mason (1986) defined the end of the Wisconsin Late Paleoindian at ca. 8000 BP, while Anderton suggested that Michigan Late Paleoindian adaptations generally extended from 9000 BP to 7000 BP (Anderton 1993:5). Based on sequences of projectile point changes nearby, the youngest Late Paleoindian Eden-Scottsbluff projectile point types fit a ca. 7630–8110 BP time range and the youngest Agate Basin types were ca. 7545–7755 BP old (Buchner and Pettipas 1990:55). The earliest copper utilization probably occurred in an initial mix with these and other Late Paleoindian stone tools. Many researchers have noticed a distinct overlap of marker artifacts from different cultural stages at many sites throughout the western Lake Superior basin (Buckmaster and Paquette 1988; Clark 1989; Salzer 1974; Steinbring 1974). It appears that the recent end of the Late Paleoindian Stage probably lies somewhere about the end of the seventh millennium BP.

Paleoindian Life on the South Shore of Lake Superior

The site 20KE20, in Keweenaw County, Michigan, is the oldest putative copper-working site in the Upper Great Lakes and in eastern North America. Sites contemporary with this ancient place are dated to well within the Late Paleoindian Stage in the region, but 20KE20, in spite of its ancient date, sits upon a land form whose age is disputed (Nordeng 1993; Pauketat 1993). There are no diagnostic projectile point styles nor raw materials typical of the region's other Late Paleoindian sites at this place, so its claim to age must instead be established by geomorphology and radiocarbon assay (table 5).

While the radiocarbon age of charcoal associated with hammered copper at 20KE20 yielded an age estimate comfortably within the range of Late Paleoindian times, the geomorphological situation proved contradictory. The site occupied a sand and gravel spit of debated age, which was composed of a series of ridges about 3.5 km west of the current Lake Superior shoreline and 300 m north of a current interior lakeshore. Now isolated, at the time in question this lake was a shallow embayment of the main water body. Baymouth sandbars developed at the modern Lake Superior level, thus closing the shallows off from the main water body. It is unclear whether the land form on which the site lies is 10,000 years old, or merely 4,000 years old.

Buckmaster and Paquette (1988) and Clark (1989) analyzed 22 artifact clusters of likely Late Paleoindian and more recent ages from the south shore of the Lake Superior basin in Marquette County, Michigan. Some included copper, which demonstrated an equivocal association of copper with the better-known and better-evidenced lithic industries and technologies practiced in Late Paleoindian times. The diversity of these artifact clusters provides a broader picture of Late Paleoindian life. The clusters occupied the edge of Deer Lake, about 20 km south of the current Lake Superior shoreline. On this lakeshore were located concentrations of prehistoric debris, which included projectile points of a material and style more commonly found in Wisconsin. They were made on a material called Hixton silicified sandstone or orthoquartzite, which comes from a known locality at Silver Mound, Wisconsin, and was widely used by people of Late Paleoindian cultures

(Porter 1961). The designs or styles of these projectile points shared much in common with others of the region, but their technical details linked them with the so-called Plano industries further to the west of the Great Lakes. Some are referred to as Eden-Scottsbluff points, and these likely date to 7000–9000 BP (Salzer 1974). Buckmaster and Paquette suggested that these projectiles were systematically deposited and ritually burned at Deer Lake as part of a human cremation, which was an early regional pattern of burial ritual in this region, judging from comparable deposits in Wisconsin, Minnesota, and western Ontario (Dawson 1983; Mason and Irwin 1960; Ritzenthaler 1972).

Elsewhere along the lakeshore were found scattered clusters of artifacts of varying ages, some diagnostic of more recent time periods. Clark (1989) remarked that although the formal tools such as projectile points were generally made of Hixton material, there were few flakes or waste fragments of this material in evidence at Deer Lake. The more mundane tools of day-in-day-out use such as scrapers, cores, and utilized flakes were made of locally available quartzites and quartz, and Deer Lake assemblages included its waste in abundance. Clark concluded that the people who occupied the lakeshore, probably over many years' time, routinely fulfilled their needs for Hixton materials by periodic trips to the quarry itself. They acquired the necessary materials for less formal tools locally, in a process not altogether different or distinguishable from other everyday activities, such as gathering food or collecting float copper.

The sites at Deer Lake appeared to be variable in age and this complicated interpretation of the role of copper in Late Paleoindian life. While Hixton material was used most heavily by earlier groups, there was at least one typologically Late Archaic projectile point that was made of Hixton, and it was found at Deer Lake site 20MQ55, where native copper was also recovered. At another locality nearby (Deer Lake #7), a Late Woodland projectile point was found in the same small assemblage as a Hixton flake. An additional two sites within the Deer Lake group included copper scrap and undated lithic debris. The association between Late Paleoindian projectile point technology and the use of worked copper was not terribly strong, so its use remained equivocal given the evidence. Clark conceded that "the Deer Lake and surrounding area may, however, be an appropriate place to investigate this issue, now that there is some baseline understanding of Plano in this part of Michigan" (110).

Late Paleoindian Adaptations in Wisconsin

In 1974 Robert Salzer reviewed Late Paleoindian settlement in the North Lakes region of the northern Wisconsin/Michigan borderlands. This area exhibited immature postglacial drainage dominated by a series of small lakes and streams and was divided somewhat by the headwaters of the Wisconsin River basin. Settlement at lakeshores and the embouchures of streams suggested sporadic occupation by small groups whose projectile points were reminiscent of Agate Basin styles cross-dated to as early as 9000 BP (Salzer 1974). The people appeared to choose warm, dry microhabitats and, like their general contemporaries, chose Hixton orthoquartzite as part of their lithic assemblage. Other local lithic materials were used as well. Salzer grouped the earliest sites into a category he termed the Late Paleoindian Flambeau Phase.

Following closely in time, the more recent Minocqua Phase also featured small sites, temporarily used by small family groups, where repair and production of projectile points took place. The projectile points were made on Hixton materials, and other tools were made of local materials. The point styles were also similar to those at Deer Lake; the Scottsbluff similarities suggested a date of 8000–7000 BP. Neither the Flambeau nor the Minocqua sites included copper.

The subsequent Squirrel River Phase included Hixton tools and copper materials. The Squirrel River site was cross-dated on the basis of similarities to the Itasca site of Minnesota at 8000–7000 BP, the same time range as the cross-dates of the Minocqua sites. Other researchers encountered contemporary variability in tool styles and adaptations throughout the western end of the Lake Superior basin (Ross 1995). The Squirrel River Phase, if cross-dated accurately, was roughly contemporary with the South Fowl Lake site, at which a developed copper-working technology existed at 6800 BP.

In another report on northwestern Wisconsin Paleoindian finds, Dudzik (1991:140) identified a scatter of projectile points, primarily surface discoveries, which lay on river terraces and portrayed the familiar Scottsbluff and Eden-like appearances noted earlier, again using Hixton materials to a noticeable degree. None of the finds was dated by other than typological means. Dudzik suggested that the northwestern Wisconsin sites represented short-term camps used by small groups of mobile hunters who captured such species as deer, beaver, and turtle, and perhaps caribou and bison (147). As elsewhere, rapidly changing environments and a mix of subsistence prospects appeared to have occurred in the region, evidenced in part by a range of lithic materials and styles, including nonlocal materials.

Early Copper-using Cultures of the Western Lake Superior Basin

From time to time, there were reports of copper artifact discoveries in the western basin of Lake Superior that resembled Paleoindian lithics, specifically lanceolate projectile points such as Agate Basin variants (Pettipas

1985; Steinbring 1966, 1975, 1991). Some researchers suggested that Late Paleoindian hunters occupied the southerly glacial/copper drift area, learned to manipulate the copper material, and essentially replicated traditional lithic projectile point forms in copper. They incorporated this novel material into a hunting tool kit to adapt to expanded northern grasslands. McCreary points occurred over a broad geographical area from southern Michigan to Saskatchewan (Steinbring 1991:27). Steinbring suggested that McCreary points were contemporary with Late Paleoindian lanceolates and were about the same age as early Archaic side-notched projectiles. Given that the Mc-Crearys were found on ancient beach ridges, Steinbring suggested that they were as old as 8500–8000 BP in western Manitoba (54–56). The source of these points, and the earliest manipulation of copper to make them, however, was easterly and southerly. According to Steinbring, "Plano populations making lanceolate stone points, especially including Agate Basin variants with a concave base, discovered copper in the Midwestern drift brought south by glaciation from Lake Superior" (56). He assumed a southern (Illinois) origin for this discovery, and an actual movement of human groups from southern environs to the expanding northern grasslands at times 500 years earlier than metal use in the Old World (57).

There were some formidable problems to be solved before such a model found wide support. The McCreary points were poorly dated surface finds and lacked tight associations with other better provenanced artifacts. The local Late Paleoindian sequence was really not well defined in time (Ross 1995:248). In fact, facing the co-occurrence of many lithic projectile points on many sites in the region, it was difficult to conclude that local complexities of cross-dating artifacts were solved well enough to advance meaningful dates for surface artifacts in isolation.

Somewhat to the west in the current state of Minnesota lay the Itasca site (Shay 1971), which included copper. It was a likely Late Paleoindian bison kill site where small groups of people occasionally camped for periods of short duration and made use of a range of local food and material resources. At this place, a small fragment of native copper, too small to determine conclusively the presence of working/hammering, was recovered from a vertically mixed context. Shay interpreted the presence of this copper material very conservatively and concluded that further excavation was necessary before it was possible to understand the place of copper, if any, in the industries of the site's Late Paleoindian inhabitants (63).

William Ross (1995) defined the earliest human use of the postglacial Minong-Agassiz Peninsula at the western end of the Lake Superior basin by the descriptive name Interlakes Composite. He defined four provisional archaeological cultures there, constituting the patterns left behind by the people who occupied the region in the millennia shortly after ice retreat. The peninsula was an ice-free stretch of land at the western end of the present Lake Superior basin, which trended northeast toward the glacial ice that occupied the north shore of the current lake.

This general region included the South Fowl Lake site, many of the sites where copper was found in uncertain association with Late Paleoindian (Plano) projectile points and other tools (Harrison et al. 1995), and many of the sites on which was found the very useful and perhaps culturally preferred Hixton orthoquartzite. The region included a very early radiocarbon date for the basin, that being an accelerator date from a disturbed cremated burial at the Cummins site at 8480 +/– 390 BP (Dawson 1983:8).

The remarkable thing about this proposed cluster of similar archaeological cultures was the great variation in the kinds of tools made and its existence at the very outset of the human use of the area. This variation appeared typical of areas further to the west (Pettipas 1985) and at least provisionally was interpreted as a normal condition in the Lake Superior basin. Ross called this phenomenon a widespread "glimpse of similar adaptations to a new land as peoples move north following the retreat of the glacial ice" (1995:256).

Late Paleoindian Summary

The earliest occupants of the Lake Superior basin were colonists in newly habitable territories made available by the changing lake extents, climatic regimes, and vegetation communities of the postglacial north. They lived in small and highly mobile groups, favoring river or lake terraces, as well as small lakes for places to live. They were extremely knowledgeable about the geographical distribution of necessary materials such as stone, as well as practiced in the skills of predicting the availability of food resources and/or animal movements. There were likely many such groups, living in small autonomous family-based communities temporarily at many locations, whose patterns of communication and interaction can be seen, archaeologically, in the heterogeneous rather light-density scattering of projectile point types and other lithic remains across the region.

The seventh millennium western Lake Superior basin sites were the homes of the earliest copper users on the continent. It is likely that they quickly incorporated raw copper from riverine and streambed drift deposits into their complex of usable material resources. Whether these users were typologically Late Paleoindian or Early Archaic in their approach to lithic technologies is really irrelevant to the problem of establishing the age of early copper use. Most researchers would agree that such evidence, though tantalizing, is not yet convincing (Clark 1995:7; Harrison et al. 1995). The best claim is put forth by the South Fowl Lake radiocarbon date on wood from the haft end of a conical copper projectile point. A developing technology in copper must have preceded it in time, and characteristics associated with Late Paleoindian and Early Archaic lithic technologies were without a doubt contemporary in this region. Some time between about 7500–6800 BP people in the Lake Superior basin began using copper metal. The South Fowl Lake date is very important because its strong context of relevance to human cultural activity makes it the reigning local standard for a reputable date. Following its example we cannot abide less relevant ones with the same degree of confidence.

The Archaic Tradition and Copper

The Archaic Tradition of 7000–2500 BP includes the prehistoric cultures from which the best evidence for Lake Superior basin copper use is found (fig. 23). However, the Archaic Tradition is very diverse and unevenly understood. The timing, variation, and boundary definition difficulties of the earliest parts of the Archaic Tradition are similar to those of the preceding Paleoindian Tradition. In addition, the unique copper tools thought to be the signature artifacts of the Late Archaic cultural stage were inconsistently dated, frequently derived from poorly documented or unknown provenance, and recovered under conditions that forever removed the best prospects for understanding them in cultural contexts. Few were professionally excavated, and those that were derived from contexts only partially representative of the whole range of prehistoric cultural activities. Burials and probable caches were well represented among excavated sites while habitation areas and collecting/hunting camps were not.

The timing of the Archaic Tradition varies from north to south, and its beginning and end are not reducible to a particular century or, in some cases, even to a particular millennium. The Archaic Tradition consists of many cultures who lived during times in which diverse local resources replaced gregarious migratory animals as preferred food. With the onset of long-lasting climatic warming trends, there were fewer large migratory herbivores to hunt. The people evolved new adaptations to supplies of locally available plants, small mammals, fish, and other resources. This emphasis accompanied a new reliance on locally procured materials for tools. In the Lake Superior region, one of those materials was copper. Though exchange of information and technologies surely continued, the elaboration of local styles was also the signature of Archaic adaptations. It is probably at this time that the spear-thrower technology, or *atl atl*, spread into the Canadian Shield environs, which one sees archaeologically as a proliferation of stemmed, corner- and side-notched projectile points. Stone tools were more frequently made of whatever local materials were available, perhaps signaling a reduction in earlier wide-ranging hunting and gathering adaptations to regional or subregional scales. The use of Hixton orthoquartzite in the Lake Superior region diminished at the margins of its earlier distribution and was replaced by quartzite, jasper taconite, and glacially derived cherts. Local adaptations included the development of functionally specific sites and technologies for gathering and hunting a broad range of foods. Fishing technologies became more important to northern foragers, based on the frequency of materials recovered and the settings of sites (Anderton 1993; Pleger 1992). A diverse diet appeared in preserved food remains at archaeological sites, especially in settings such as rock shelters where preservation of organic remains was enhanced (Wittry 1959).

The regional Archaic Tradition is best known from the excavation of prehistoric mortuary sites. The resultant archaeological literature emphasized the definition and differentiation of three overlapping archaeological complexes of the Middle and Late Archaic Stages: the Old Copper Culture (OCC), the Glacial Kame Culture (GKC), and the Red Ochre Culture (ROC). These complexes are best known from excavations of burial localities that sometimes included evidence of mortuary ceremonies, with grave goods of copper, red pigments, shells, and ornate stone tools. These manifestations seemed similar enough over wide ranges of time and space to suggest that they were somehow linked together. For the last half century, archaeologists have struggled to understand the chronologies, relationships, and meanings of these three complexes.

The Early Archaic Stage (7000–5000 BP)

During the Early Archaic Stage, gradual alteration of site locations, subsistence practices, technologies, and other aspects of cultural adaptations to new configurations occurred. These adaptations lasted thousands of years and appeared to be comfortably in balance with local resources. The transition was so uneven in time and space that neither the Early nor the Middle Archaic Stages is well known from the region. In fact they are barely a presence on Isle Royale (Clark 1995), despite radiocarbon dates from copper-mining pits late in the time range. Salzer (1974) suggested that the transition to the Archaic Tradition was gradual and poorly represented in the archaeological record. He defined the North Lakes Squirrel River Phase at an estimated 8000–7000 BP age and discovered that site locations replicated those of the earlier occupants of the region: a stream outlet at an interior lake. Age estimates for the phase were derived from typological comparisons with other tool complexes. The tools and other materials recovered shared functional similarities with those from the Itasca site of Minnesota, but included the presence of conical copper projectile points and other pieces of worked copper (Salzer 1974:45). Stone tools such as scrapers were made of both exotic and local lithic types. Very little was known of the subsistence practices of these people, for no food remains were reported. Salzer remarked that few sites of comparable age were discovered in the North Lakes region. Stoltman proposed that research into the characteristics of the ancient environment offered the best prospect for inferring the probable life ways of these people, because few subsistence remains were actually recovered from their camps (Stoltman 1986:215).

Warmer, drier conditions than those of today prevailed. These conditions provoked shifts in vegetation zones, forest composition, and animal habitats, with extensions of southerly animals likely reaching further north than at present. The north and the south of the region were distinctly different. Dry grassland conditions and oak-dominated forests occupied southern Wisconsin and Michigan, while mixed deciduous and pine forests occupied the north. In both areas, small bands of people probably trapped, fished, and hunted elk, deer, small mammals, and seasonally abundant fish. Plants and storable nuts probably provided food security, though the abundance and particular species of interest no doubt varied over the area. The rarity of known sites in the general time range, especially adjacent to major lakeshores and strand lines, implied (1) a drop in human population levels from earlier times, (2) an inability of archaeologists to identify accurately the remains of Early Archaic adaptations, and (3) a loss of sites from the archaeological record due to changing lake levels and river gradients, or a combination of these factors (217).

The Middle Archaic Stage (5000–3000 BP)

The people of the southern Lake Superior basin left behind a more visible archaeological record beginning about five thousand years ago (table 6). Their presence included mining pits, habitation sites, fishing extraction sites, workshops for copper tool fabrication, quarry sites, stone tool preparation areas, caches of tools, and cemeteries.

Radiocarbon assays from mining pits on Isle Royale yielded a consistent assessment of the age of the fill to the fifth millennium before present. Clusters of dates elsewhere supported the antiquity of copper technology to times prior to the Lake Nipissing maximum (Stoltman 1986:225).[3] Archaeological evidence recovered in situ suggested the proliferation of copper-

Table 6. Some Middle Archaic dates for copper from Wisconsin and northern Michigan.

Site/Citation	Lab ID	RC dates (BC)
Oconto (WI) (Stoltman 1986; T. C. Pleger, personal communication, 1996)	C-837/839 C-836 GAK Pleger AMS (unpub) Pleger AMS (unpub)	5560+/–600 3650+/–600 2590+/–400 4070 3300
Minong (Isle Royale) (Clark 1995; calibrated intercepts in parentheses) Lookout (Isle Royale) (Clark 1995)	M-1384 M-1390 M-371e W-291 M-1388 M-1385 M-1275 d,e,f,g	3510 (3040) 2626 3500 (3034) 2618 3629 (2278, 2234, 2209) 933 2135 (1614) 1100 2140 (1851, 1850, 1761) 1463 2019 (1677) 1410 3023 (2857, 2821, 2691, 2689, 2660, 2637, 2623) 2330
Chautauqua Grounds (WI) (Pleger 1992)	TO-3983	2760+/–150

working technology at this time. Tool production for such tasks as food procurement, food processing, and the processing of materials such as hides and wood were found. In addition, the production of ornaments and the elaboration of their shapes, functions, and means of attachment occurred. At several of the excavated cemeteries, artifacts of all kinds were more frequently encountered in association with children and subadults than with adults. Despite this fact, several male adults were interred with substantial copper, and women were buried with it, too. Little is known about the particular social and ideological beliefs that propelled mortuary behaviors. It is likely that each cemetery represented multiple cultures, at least separated in time, and that more than one sort of mortuary behavior took place there. Thus children owned highly individualized items in death, just as male stand-outs in life owned them. It is clear that both were buried with copper possessions, perhaps for different reasons.

Artifacts from excavated habitation sites yielded inferences about subsistence practices. The location of the Chautauqua Grounds site and its specialized tool inventory related to fishing was one of the most informative of Middle Archaic sites (Pleger 1992). Stoltman suggested that southern and western Wisconsin's rock shelters may have been the winter homes of Middle Archaic peoples, who for some reason did not use or leave their copper tools there (Stoltman 1986:226). Strong similarities between noncopper artifacts found in both geographical areas substantiated this claim. In these rock shelters were found food remains such as deer, small mammals, and birds. Archaeologically recovered data suggested seasonally changing subsistence regimes and seasonal movement among a range of ecological settings, a pattern that probably began in Late Paleoindian times, involving high levels of social interaction among small forager communities. It is reasonable to assume that over the spatial extent of copper tool use and the time range allotted to the Middle Archaic Stage, many variations in cultural behavior existed. This would be consonant with the variation that is apparent in Middle Archaic mortuary behavior.

What Was the Old Copper Complex?

Was it a series of related archaeological cultures, simply a descriptive device useful only to archaeologists to rescue their data from chaos? Was it a loose category of convenience for all copper artifacts from a particular region found without full provenance? Was it a romantic notion of a lost civilization adhered to by those willing to adopt simplifications about the variability inherent in past societies and their adaptations? Was it a viable category of social meaning and identity in the minds of those people who made the artifacts and disposed of them in caches, burials, and elsewhere? The nature of the Old Copper Complex (OCC) was debated by archaeologists for years (Binford 1962; Clark 1995; Fogel 1963; Mason 1981; Stoltman 1986). Most agree now that the OCC included a range of societies or communities interacting through a complex of beliefs demonstrated by mortuary customs. In addition groups of people shared a technology or technologies related to copper tool manufacture, and they were linked across vast reaches of time and environs by means of communication systems that may have been affirmed by hand-to-hand trade.

The core artifacts thought to represent the OCC occurred with greatest frequency in eastern Wisconsin and in adjacent areas of the Upper Peninsula of Michigan. The classic collections of these artifacts were made by nineteenth-century farmers as well as by men of science. There were also roving entrepreneurs who sometimes paid good money for their acquisitions; the stories of these early collectors' adventures make great tales themselves (Penman 1977:5). The interest in copper tools also generated a nineteenth-century cottage industry of fakery, another interesting but tangential tale. There was no doubt some causal relationship between the concentration of the artifacts' provenance and the degree to which accelerating human activities such as agricultural clearing, logging, log driving/milling, and power generation exposed older buried strata.[4]

Some of the artifact types associated with the OCC occurred in widely disparate places and times within eastern U.S. prehistory, so it was problematic to call them OCC cultural markers. For instance, copper awls and beads were widespread artifact types. Copper crescents occurred in Initial (Middle) Woodland contexts in Upper Michigan (S. Martin 1993) and in Ohio Hopewell settings (John R. Halsey, personal communication, 1996). Nonetheless, as Mason pointed out, there remained a group of unique tools bounded in a regional sense and apparently in time (Mason 1981:188). He identified this subset of copper implements to include socketed projectile points and knives, socketed axes or adzes, spuds, and semilunar knives. Many of these implements occurred as surface finds with their greatest known frequency in eastern Wisconsin. In addition they were excavated from burial contexts at three well-known localities across the state.

The Osceola site, in Grant County, Wisconsin, was one of the earliest excavated Old Copper Complex sites. Osceola was one of three Wisconsin cemeteries representing cultures that interred their members with a range of copper grave goods. Yet these places represented several millennia of use; the assumptions that their makers belonged to a single society, or even a single archaeological culture, was unrealistic. The Osceola site, excavated by the Milwaukee Public Museum, represented one of the first in situ discoveries of the classic complex of Old Copper artifacts (Ritzenthaler "Osceola" 1957:187). Despite its geographical location peripheral to the Lake Superior basin, the site must be discussed in the context of the Old Copper Complex to understand fully the history of and meaning behind that concept.

The Osceola site consisted of a burial pit that was partially eroded and destroyed by the Mississippi River. Thus the size of the site before disturbance was not known. What remained was at least 20 x 6 m in extent and as deep as 1.5 m below surface. Copper materials were occasionally associated with bundle burials in the bottom half of the pit, but, according to Ritzenthaler, were "in most cases" scattered within the general fill (188). The general fill, presumably borrowed from nearby, also included materials from a Woodland-age culture and provided a cross-date for the establishment of some of the pit fill.

An estimated 500 people were buried at Osceola. They were generally secondary or reburied bundles, or, in some cases, partial cremations. "Osceola artifacts did not occur directly with the burials, but were in all cases in the same stratum" (197). Based on the variety of interment styles and conditions of preservation, the researchers concluded that the Osceola site represented a lengthy duration of use. They suggested that the Osceola, or older component, lay somewhat underneath and adjacent to a more recent Woodland-age component. The copper artifacts included awls, spuds, socketed spear points, knives, and socketed conical points. The other copper materials probably functioned as ornaments (193). Exotic materials, or materials from nonlocal sources, included red ochre (hematite) and galena (lead) cubes.

Initially, researchers thought that the Osceola component represented a pre-pottery Woodland Tradition site. Later, radiocarbon assay of the OCC component run by the University of Michigan yielded a date of 1500 B.C. +/– 125 (M-643). Yet Woodland affiliations were noted for virtually all data classes. Old Copper materials were explained by simply suggesting that the OCC persisted until Woodland times and overlapped the boundary between the Archaic Tradition and what followed (202). This conclusion seemed valid, because continuities among site locations, artifact types, and burial modes were the rule rather than the exception elsewhere in the OCC region.

The second puzzling OCC locality, the Oconto site (Oconto County, Wisconsin) yielded an initial radiocarbon assay of unexpected antiquity, in fact somewhat earlier than the agreed-upon dates for the beginnings of the Middle Archaic Stage. The dates, ca. 7510 BP (C-837 and C-839) and 5600 BP (C-836), were considered both on typological and geomorphological grounds to be in error and were originally rejected. Acknowledging the experimental character of radiocarbon methodology at the time, everyone believed that the dates were at fault. The site was important because it, like Osceola, constituted one of the earliest OCC manifestations to be excavated, analyzed, and dated by professional archaeologists. Reexamination of Oconto materials by means of AMS dating of organics associated with copper artifacts again suggested an antiquity of ca. 5300–6000 BP (Thomas C. Pleger, personal communication, 1996).

At Oconto, test excavations by the Oconto County Historical Society, the Milwaukee Public Museum, and the State Historical Society of Wisconsin revealed human burials accompanied by copper artifacts (Ritzenthaler and Wittry 1957:223–24). The site consisted of at least 29 burial and cremation pits in gravel, filled with sand, ten of which included artifacts. Earlier disturbances probably destroyed a portion of the site. Extended burials, multiple burials including bundles, and single/multiple cremations established that multiple styles and numbers of interments were present (230). The placements of individual pits avoided earlier interments, which suggested that variations in burial mode did not represent a time sequence.

A total of 26 copper implements was recorded: "Copper implements, particularly awls, occurred sporadically in the upper levels of the sand layer and bore no apparent relationship to the burials" (231). Two cremation pits contained a single artifact each of copper. Awls and crescents constituted the most numerous artifact classes; three spear points, a fishhook, and number of unidentified pieces completed the collection. The authors noted the lack of ornamental copper items in the small collection. Whelk shell of Atlantic coast origin and iron ore were also recorded; it was assumed that the iron originated somewhere near the site.

At the time of the original publication, which preceded the announcement of the radiocarbon assay results, site investigators concluded that Oconto was contemporary with Osceola and that both represented the OCC. It is now reasonable to question that both sites are the same age (Pleger, personal communication, 1996). The dates represented a 5,000-year time range, calling into question the closeness of any cultural affiliation between the sites. The burial modes were actually quite different; those at Oconto were individual pits with many extended inhumations as well as cremations. Those at Osceola resembled a Woodland ossuary of secondarily deposited bundles, though its researchers concluded that the burials occurred in increments rather than all at one time. At Oconto, eleven copper artifacts (crescents, spear points, bracelet, spatula, fishhook awl) were immediately associated with burials or cremations. Overall, artifacts of all types were found more frequently with nonadults than with adults. Preserved fibers associated with copper artifacts were the sources of the 1996 AMS dates from Oconto, which repeated and strengthened claims to a Middle Archaic age. It is difficult to extend the age of the Osceola site to 6000 BP solely on the claims to similarity with Oconto. Osceola and Oconto represented vastly different cultures, environments, and time spans. Whether Oconto's users were ancestral to the peoples of the Osceola locality is unknown.

The final member of the classic OCC triad was the Reigh site (Winnebago County, Wisconsin), excavated by the State Historical Museum, the University of Wisconsin, the Oshkosh Public Museum, and with the support of the Wisconsin Archeological Survey. The Reigh burial complex appeared to predate an overlying pottery-bearing component. Fourteen burial units were recovered during professional excavations, containing the remains of 44 people. Additional burials were apparently excavated at other times by collectors (Ritzenthaler, "Reigh" 1957). Some of the burials were superimposed upon one another. As at Oconto, multiple burials in mixed modes were common; as at Osceola, the burials were clustered in a rather defined space. One adult male was interred with what appeared to be a copper headdress, and was accompanied by at least four other individuals, including children. "Knives, hematite pebbles and copper points were associated with and presumably used by both males and females" (Baerreis, Daifuku, and Lundsted 1957:273). The homogeneity of the artifact assemblage, the lack of a patterned sequence of interment styles that could be linked with a chronology, and the location of all of the burials within a rather tightly defined space suggested to the authors that the materials represented a single cultural complex. These people used the place over enough time that the specific locations of earlier burials were disregarded in the course of finding places for newer ones. The researchers found closer parallels between Reigh and Oconto than with Osceola. Multiple modes of burial, copper implements and ornaments, the discovery of a marine shell gorget, and dog burials suggested a link with the Glacial Kame (GKC) mortuary complex as well (276). The final estimate of the age and affiliation of the Reigh placed it within the Early Woodland Stage as an example of the Glacial Kame mortuary complex, with holdovers from the OCC of earlier times, a "cultural bridge between Old Copper and Early Woodland (Ritzenthaler, "Reigh" 1957:278). A radiocarbon date from the Reigh site suggested an age of ca. 3660 +/− 125 BP (M-644, uncal).[5] Numbers of so-called classic OCC artifacts were recovered from additional excavations at the Reigh site (282–310) including celts or wedges, crescents, spuds, and socketed or stemmed projectile points and knives. The discovery of these artifacts supported the restoration of at least portions of the Reigh site to an OCC affiliation.

Wittry and Ritzenthaler's summary statement about this triad of important sites was reprinted in 1957 in *The Wisconsin Archeologist*. They concluded that similarities among the triad lay both with "the occurrence of Old Copper implements at both of the sites and certain similarities in the burial patterns" (1957:321). They identified the Osceola site, the outlier to the general OCC area of distribution, as more recent than Oconto, and suggested that the rolled socket copper projectile type predated the angular ridge-backed type. Part of their hesitancy to accept the very old dates from Oconto must have been the lack of any overall archaeological chronology for the region.[6] Their assessment of the origin and spread of the OCC placed Wisconsin in the center of its development: "We postulate that the origin of the Old Copper complex occurred when an early hunting and gathering group living in the Wisconsin area began to utilize native copper for the production of the distinctive utilitarian types of that area. At first, the nuggets in the glacial drift provided a handy source; later, copper was quarried from the trap rocks of Isle Royale and the Keweenaw Peninsula" (323). This scenario would probably be agreed upon widely even today. The earliest known classic OCC artifacts do come from Wisconsin, and they are to be found at the Oconto site.

But what the OCC represented in terms of human institutions and social interactions was still elusive. In one interpretation, the OCC was a local version of a far more general and widespread complex of related mortuary activities. Within this complex the co-occurrence of copper and other materials such as animal bone whistles, marine and local shell ornaments, iron ore and pigments, and finely worked projectile points and blades was prominent. It was found in various combinations at many prehistoric cemeteries throughout northeastern North America, especially in (roughly) the 4000–2000 BP time range (Fogel 1963; Hruska 1967; Pleger 1996).

Within this complex, some individuals announced their social worth, economic connections, and kin relationships through differential inclusion and display of copper artifacts in burial accompaniments (Binford 1962).

In other views, the OCC was an example of how people made use of whatever materials their local environs had to offer. In this material and technological view, copper was regarded as a material that performed as a fortuitous substitute for the dearth of workable stone in the core area (Clark 1995). This viewpoint was somewhat supported by the observation that virtually every implement made of copper had a lithic or bone counterpart in other prehistoric assemblages in and outside of the copper-rich area. In this view, the particular tools of the OCC (Mason 1981; Stoltman 1986:217) that occurred in relatively fixed frames of time and space constituted a technological complex related to subsistence tasks and woodworking, which other contemporary peoples practiced using tools of other materials. Clearly, these two views were not mutually exclusive. It is perfectly reasonable to suggest that copper played technical, social, and ideological roles in highly variable ways throughout prehistory and that this variability was exposed by an examination of the three classic Wisconsin cemeteries.

New research at the Oconto site and at the Riverside site in Menominee, Michigan, and examination of comparative collections from other northern copper-bearing sites such as the Chautauqua Grounds site clarified the links between the varied mortuary complexes of the Middle and Late Archaic Stages (Pleger 1996). These reanalyses considered the human adaptive behaviors (such as subsistence practices) that remained uninvestigated, and incorporated improved dating techniques. The Chautauqua Grounds site in Marinette, Wisconsin, was critically important for developing new understandings about lakeshore subsistence practices of Middle Archaic peoples. There were two important findings relating to copper from this complex of sites. First, the site lay adjacent to a productive warm weather fishery that provided an abundance of easily obtained fish. Such a setting was doubtless appealing for people of Woodland times, whose cultural remains were found in profusion there. But the fishery was already an important resort for food for people of earlier times. At Chautauqua Grounds, an AMS radiocarbon date on wood from the interior of a copper conical projectile point confirmed the existence of both a developed copper-working technology and a specialized fishing adaptation at 4210 +/– 60 BP (Pleger 1992:174). Pleger noted that the copper assemblage at Chautauqua Grounds, which was of mixed age, shared functional and technological similarities with those from Oconto (35 km distant) and the Riverside site (Hruska 1967), about 7 km upstream from the lake, despite vast differences in their absolute ages.

The Late Archaic Stage (3000–2000 BP)

The classic site whose excavation linked the Late Archaic Stage with earlier cultures was the Riverside site, excavated by the University of Michigan Museum, the Oshkosh Public Museum, and the Milwaukee Public Museum (Hruska 1967). Subsequent research by Thomas Pleger (1996; Pleger, personal communication, 1997), including the generation of a new series of AMS radiocarbon dates, verified the use of the Riverside site from ca. 1000–400 B.C. As with many sites in the region, people used the site through long reaches of time; the Riverside site included a Woodland component, which dated to ca. 650 A.D. The site lay near the Menominee River on a high stable sand dune. Burial pits and village remains there apparently represented the activities of a variety of archaeological cultures over many centuries.

What attracted attention first was the cemetery at the site, which was reminiscent of the Wisconsin sites described earlier. At Riverside, 80 features were excavated, most of which were single and multiple interments. Burial modes were highly variable and included ochre, copper, hornstone, and other exotics such as obsidian. Preservation of organic materials was enhanced by contact with copper oxidation; in some cases fabrics or textiles, rarely discovered in the Upper Great Lakes, were preserved (King 1968).

Of the 80 pit features investigated, about 54 included human remains or probable mortuary evidence. When corrected for cases of multiple interments, 14 of 54 interment units included copper artifacts. In 7 of these 14 cases, copper accompanied the burials of children or subadults. These burials were notable for the variety of copper artifacts included. They were generally implements, or what have been classed as implements, rather than ornaments. Celts, awls, crescents, and projectile points occurred in burial contexts with children; in two cases the interments were multiple and include adults with children or subadults. At least one child was buried without copper but with more than 100 lithic blades (feature 63). One adult female (feature 27) had a lavish burial (ochre, shell beads, hornstone) and one adult female (feature 41) died a violent death (Hruska 1967:212). The University of Michigan excavations documented a female and child burial (feature 14), which included obsidian, copper beads, and hematite (red ochre), dating to ca. 800 B.C. (Thomas C. Pleger, personal communication, 1997).

Being buried with copper artifacts was not something that everyone did or had done to them; only about 24 of 63 people were buried with actual copper artifacts. Relatively lavish inclusions were most likely to accompany multiple burials when one of the interees was a child or subadult. Some researchers found the inclusions of copper and other exotics in disproportionate amounts with fetuses, infants, children, and women to beg explanation. If status in life were to be reflected in death by numbers of copper artifacts, then one must explain why the young and the female had the preponderance of recognition. Another interpretation suggested that something other than, or in addition to, activity in life was being stated by these grave inclusions. The association of copper with children and subadults may have had ideological causes; the evidence of this link is another matter. Some native belief systems of the sixteenth through nineteenth centuries in the region regarded copper with awe, due to its association with supernatural creatures such as the lynx and/or serpent manitous. Possession of copper was thought to be associated with health, wealth, and well-being (Barnouw 1977; Hamell 1987). Others suggested that the inclusions of copper artifacts and/or exotics such as obsidian, shell, and hornstone demarked families who actively engaged in trading partnerships or institutions (Pleger 1996). Obviously, trading links/mechanisms and the motivations and ideological rationales for them can/do co-exist, so the two explanations were not necessarily exclusive. In fact, they appeared to support one another.

Pleger identified the Riverside cemetery as an example of the Red Ochre Complex (ROC), a mortuary complex more recent in time than the OCC, but perhaps related. The ROC lasted from ca. 1200 B.C. to about A.D. 1 and linked the local cultural behaviors of the Archaic peoples with their Woodland successors. Both sturgeon fishing and the use of wild rice were possible subsistence strategies for ROC people and their contemporaries (Ford and Brose 1975; Hruska 1967). Pleger summarized the ROC adaptations and mortuary customs: "The geographical placement and physical setting of Red Ochre cemeteries suggests the presence of multiple hunter-gatherer group territories settled within riverine and lake systems within the western Great Lakes" (1996:9). The presence of exotic materials that originated outside the Lake Superior copper district, such as hornstone from Indiana/Illinois, shell from the Atlantic and/or Gulf coast, and obsidian from the western United States (Griffin 1965) implied widespread communication among these groups. There were strong similarities between the archaeological record of these groups and their contemporaries north of the Lake Superior basin (Wright 1995:290). Pleger and others linked the presence of exotic materials with the emergence and maintenance of trading links that ultimately remade the social relations of local people in levels of increasing complexity, particularly social differentiation. This trend linked the western Great Lakes with developments to the east and south at the beginning of the first millennium A.D.

Archaic Stage Life: Lake Superior and Isle Royale

The habitation sites of Archaic times are not well known, but they are essential for reconstructing daily life in the past. They are found in the densely forested and rugged terrain in the Upper Peninsula of Michigan and adjacent regions of Wisconsin, as well as on Isle Royale. Recent research in these places interpreted Archaic life in the copper-bearing region. Despite the proclamations of ancient use of Isle Royale for copper extraction, its early archaeological record is not as ancient as that of the mainland. It began about 4400 BP, according to the well-developed radiocarbon sequences on the island. The Early and Middle Archaic Stages were barely represented there, and the classic implements that defined the so-called Old Copper Complex were not there either. Some of these data gaps were due to research skews, and the limited sample of discovered sites probably was not representative of the whole island (Clark 1995).

Clark identified the earliest inhabitants of the island by their apparent lithic preferences and their preferred site locations. Their sites were found at elevations between 195 m and 201 m, reflecting higher lake stages in Archaic times. They made exclusive use of Portage Lake quartzites and rhyolitic lithics (162). Radiocarbon assays in the 3500 BP time range bore out these assessments. Such tendencies (lithics, locations, and dates) made it possible to identify putative Archaic sites without confusing them with aceramic Woodland (more recent) sites. Clark linked the earliest users of Isle Royale's copper resources with people from the north shore of Lake Superior, whose archaeological sites and assemblages shared similarities with those of the island. These were identified as part of the Shield Archaic archaeological culture (166). Despite known sampling skews, those people using Isle Royale in the fifth to third millennium were members of cultural groups distinct from those of the south shore of the lake basin.

On the mainland there were interesting habitation data from small camps assumed to be contemporary with Late Archaic cemeteries elsewhere. Their humbleness was perhaps their most interesting characteristic; they repeated the familiar pattern of long-lived reuse of the same localities through hundreds of years or more. Sites in the North Lakes (Wisconsin) region, in the Deer Lake region of Michigan, in the interior of the western Upper Peninsula, and on the Keweenaw Peninsula represented what was known of Upper Great Lakes Late Archaic life away from the vivid mortuary contexts of the interior and south. As an example, the site 20KE20 represented a habitation and copper-working site that was used sporadically at least 3,300 years ago (S. Martin 1993:175). Its advantageous location vis à vis food and subsistence, shelter, and copper procurement probably encouraged reuse. Occupation and copper working took place on a series of beach ridges and probable lake-level regression features overlooking a small interior lake. These ridges yielded a radiocarbon sequence from 7800–1600 BP (175).

Many artifact collectors visited 20KE20 prior to formal excavation, and discovered copper artifacts using metal detectors. Whole classes of artifacts were missing from the excavated subassemblage as a result. Most of the casual finds came from an 192–195 m beach ridge elevation and were typologically younger than those expected from a bona fide OCC site, but fit well with the ca. 3300 BP radiocarbon dates there. It is unclear whether the core group of OCC tools were ever collected from the place. Most materials that remained after casual collecting were artifact types that were either chronologically insensitive or were thought to represent younger archaeological cultures. These finds included, in the first instance, beads, awls, conical points, wedges or blanks, and needles, and in the second, so-called butter knives and hairpipes. The only classic OCC tools represented in the known collections from the site were crescents and socketed projectile points similar to Wittry Type Ib, but lacking rivet holes (Wittry 1957).

The stone tool subassemblage from 20KE20 was chronologically insensitive. It consisted of small amounts of quartz, quartzite, basalt, and chert debitage, representing leftovers from the probable manufacture of bifacial tools on flakes. Very few of the lithics recovered were finished tools. Quartz was the most common material and appeared in uneven clusters in some areas of the site, probably related to areas where cobbles were hammered to produce flakes. In all, the lithic subassemblage was small and appeared (1) relatively unimportant compared to the copper subassemblage, (2) related to very light use of the area by small, seasonally resident bands of people, or (3) both (Pauketat 1993:180).

Four probable hearth features were excavated; they contained fire-cracked rock and reddened ashy soils, and occasionally fragments of worked copper and/or charcoal. No food remains were recovered. The distribution of hearths and their contents again suggested light, periodic, specialized use of the site locality for small groups of intermittent visitors, rather than sustained use.

The Deer Lake sites of Marquette County indicated a faint Late Archaic presence at the shorelines of small interior lakes. These sites documented a shift in preferred lithic materials away from Hixton orthoquartzite to quartzite and then to Hudson Bay lowland cherts and quartz, particularly for the production of small and expedient tools such as scrapers (Clark 1989:110). The Archaic components were found superimposed on or adjacent to earlier Plano materials. The recovery of Bayport chert at site 20MQ47, a chert type from the Thumb of Michigan's lower peninsula used widely from Late Archaic times onward, suggested that the users of the site engaged in intermittent travel or trade southward. At site 20MQ55, a probable Hudson Bay lowland chert projectile point known to be of general Late Archaic provenance was found with a tanged point of native copper. Native copper in association with nondiagnostic lithic assemblages were also discovered at sites 20MQ56 and 20MQ57. Some of the quartz or quartzite debitage at these places probably was of Late Archaic age. Local copper was no doubt an important entity in the array of resources at these and other Lake Superior basin sites.

At the western end of the Upper Peninsula, rough terrain and a focus on copper procurement sites deterred, to some degree, the investigation of Late Archaic habitations and lithic quarry sites. But on the Ottawa National Forest, the discovery and excavation of the Ottawa North and Alligator Eye sites enriched an archaeology-based understanding of Late Archaic daily life. Both sites were found south of the copper-bearing trap range, in the midst of a complex network of small lakes. At the time of their occupation, the general area of the sites was covered by mixed hardwood-conifer forests with pine inclusions in a situation that would have been exposed to winter weather. Hill suggested that the sites fit into a wider pattern of resource use and population movement, in which small groups of people who moved north in the summer returned to the south periodically to spend the cold seasons (1994:46). This linked them to related groups of people, perhaps family members, at mortuary sites such as those represented by the Riverside site. Resource exchanges, such as copper and Hixton orthoquartzite, were logical within this model and their archaeological distributions across the major regional biotic divide lent support to their operation.

The Ottawa North site was a small, one-time occupation and one of the very few Archaic sites in the region at which food remains of small mammals were found. The remains of *Marmota monax* (woodchuck) and another unidentified species of small animal were recovered from a hearth feature there. A radiocarbon assay of this feature dated it to 3220 +/– 220 BP (uncal) (Beta-42451; Hill 1994:27). Locally available quartz made up most of the stone tool subassemblage. Cobbles were likely reduced at the site to provide expedient tools for cutting and other functions. The spatial distribution of lithic debitage suggested to Hill that food preparation activities were centered at the hearth feature, while expedient tool production took place a few meters away. This pattern was consistent with the use of the site by a small family group during a warm-season short-term occupation.

People roughly contemporary with those at Ottawa North may have resorted to outcrops of quartz within their home ranges to gather flakes of material from which to produce large formal tools such as projectile points. One such place was the Alligator Eye site, a quartz quarry researched by the staff of the Ottawa National Forest (Hill 1994). The site may have been mined by fire setting and intentional facial cracking, in a pattern that was reminiscent of the alleged traditional means of copper quarrying. Both expedient and formal tools were common at the site, with functional types such as spokeshaves, bifaces, projectile points, gravers, unifaces, and woodworking tools recovered. Radiocarbon assays dated the site's use to ca. 3640 +/– 150 and 3490 +/– 110 (uncal) BP. The predominance of quartz shatter and debitage suggested that the site was heavily used for warm season tool production and perhaps for preparation of organic materials such as wooden tool handles.

In the adjacent North Lakes district of Wisconsin, Late Archaic remains were hard to find because the sites were small and apparently lightly used. Salzer defined the Burnt Rollways Phase at three components within the region. The type site sat on a stream bank and included a well-made quartz and exotic chert industry in addition to a developed copper industry featuring tools such as fishhooks and awls. Salzer estimated that the communities responsible for this evidence lived in the area about 3,000 to 4,000 years ago (Salzer 1974:47), dates roughly contemporary with the people from 20KE20 and the Ottawa North sites. Recent work at stratified sites in the Wisconsin headwaters area yielded evidence of on-site copper working and subsistence remains such as turtle, beaver, and blueberry seeds in a Late Archaic context. Similar in assemblage contents to the Burnt Rollways Phase, the radiocarbon assays dated the components to ca. 3200–3600 BP (Moffat and Speth 1996:28).

The End of the Archaic Stage

What is known and unknown about the Archaic Stage and its cultural adaptations? First, we still know very little about the particulars of the many local differences in subsistence and settlement practices, population levels, and the full range of technologies used in everyday life. In situ data are weak or poorly preserved for these information classes, and sites are small and lightly used and perhaps not terribly numerous. We cannot yet track the development of the copper-working technologies of the Middle and Late Archaic Stages. Discerning evolutionary sequences in technology is made all the more difficult by the provenance uncertainties of many of the artifacts we do have to study. Biases in the Archaic data base are clearly apparent but unmeasurable.

Aside from the fact that the mortuary data show some levels of interrelationship among the peoples and ideologies of the Middle and Late Archaic periods, we cannot yet account for the means by which they communicated their cultural similarities and differences to one another. We are not sure of the relative ages of the mortuary sites, of how persistently they were used, or of how large they once were. Nor can we account, beyond the hypothetical, for the means by which some people acquired and retired from life with exotic materials such as obsidian and marine shell. We can only hypothesize about the degree and mechanisms of their economic, social, and ideological links to neighboring people and whether these relations are linked to the development of systems of social differentiation at home.

The Initial Woodland Stage

The appearance of ceramics marked the archaeological distinction in Late Archaic and Initial Woodland life on sites in the Upper Great Lakes region (fig. 24). This change occurred about the middle of the third millennium BP. By 2000 BP pottery was well established in the region. Where the idea of pottery originated and how that idea spread into the basin is not clearly understood. The earliest pottery in the region, called Dane Incised, from southern Wisconsin, was similar to pottery that was found further to the south in the large river valleys of eastern North America. Pottery likely spread slowly north and perhaps east until by the middle of the third millennium BP it became part of waning Late Archaic culture. Other Woodland characteristics included the spread of cultigens and the development of artificial burial mounds. These traits became part of Woodland adaptations later in the Upper Great Lakes region than elsewhere. Exotic cultigens were not important to Initial Woodland adaptations to the Lake Superior basin, so far as is known (Stoltman 1986:273). Burial mounds extended from the Rainy River district of Manitoba to the west of the basin. It is possible that this western region was the source of the important Laurel pottery that spread into the lake basin. The Initial Woodland people were most closely linked with peoples who produced Laurel pottery, rather than with peoples to the south of the Great Lakes. Nonetheless, some of the Initial Woodland groups were seriously occupied with copper-working technologies. Copper was without a doubt important to peoples to the south, where it became part of Hopewell mortuary inclusions. Researchers continue to investigate the organization of trading relationships with the Hopewell peoples of Illinois, Indiana, and Ohio (279).

The archaeological sites of this time period and cultural affinity tended to be stratified with components of other time periods. Most sites were low-impact habitation and resource-gathering places used repeatedly by small groups of people. Though there was widespread similarity in the sites across the Lake Superior region, the preservation of sites in a range of environments tells us much about the people and their daily activities. There was clear continuity with prior times in the copper-working technologies of the Initial

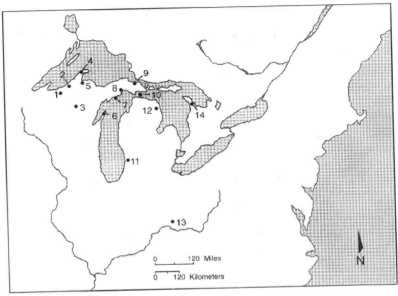

Fig. 24. Woodland Tradition sites and sites of indeterminate age: 1—Timid Mink, 2—Minesota/Mass-Caledonia, 3—Robinson, 4—Waterbury/Delaware, 5—Sand Point, 6—Richter, 7—Summer Island, 8—Winter, 9—Naomikong Point, 10—Juntunen, 11—Dumaw Creek, 12—Thunder Bay, 13—Hopewell/Mound City, 14—Hunter's Point. (Drawing by Brett A. Huntzinger.)

Woodland peoples, but it was more prominent and perhaps more important on some sites than others.[7]

House Life among Local Initial Woodland Foragers

There are solid data about Initial Woodland community organization and the physical structures in which people lived. These data come from a number of sites and present a nice range of contrasts. Some are major lakeshore settings and others are found at interior lakeshores. Together they represent some of the range of possibilities available in the Initial Woodland housing situation. More to the point for prehistory, they probably represent two seasonal alternatives for house construction, community size, and access to food resources.

At Summer Island on the north shore of Lake Michigan, David Brose excavated a small Initial Woodland camp about 1,800 years old. The site included excellent faunal preservation; seven species of fishes, eight species of mammals, three species of birds, and one species of turtle was recovered. This array suggested a broad-spectrum diet with seasonal (summer) emphasis on fish, small fruits, and small animals. Brose interpreted the remains to represent seasonal use of the island during warm seasons by small groups of foragers. The spring sturgeon fishery was particularly important (Brose 1970:148). The pottery left behind was similar to other Laurel pottery, found from the St. Lawrence River to Lake Winnipeg (215). The bone and ground stone industries featured artifact types that were used in fishing and making nets. Brose estimated that the community population was about ca. 25–40 individuals.

The community consisted of about four oval houses, hearths, drying racks, storage pits, and refuse areas. The houses were framed by a series of large (ca. 20 cm diameter) and small (ca. 10 cm diameter) posts in the ground. The houses measured about 7–9 m long and from ca. 4–5 m feet wide. Each oval house had one or two hearths inside. The floors were hard-packed sand, dirt, and midden (undifferentiated organically-rich compressed sheets of disturbed soils). Brose suggested that the larger structures housed extended families around which perhaps more than one woman did her cooking and household chores. Brose reasoned that the houses stood for about three to four years before decaying. Historic accounts of native houses in the region, albeit 1,000 years more recent than the Initial Woodland, described oval, dome-shaped frames of lashed poles over which sheets of bark were laid as sheathing. Perhaps these were similar to the prehistoric houses.

The Summer Island people worked and used copper, and made a range of implement and ornament types. Tools appeared to be less brittle than ornaments, indicating that annealing was not the final manufacturing stage for ornaments. This finding duplicated that of Leader (1988), which suggested that tools broke less often if well annealed. The tool variability and the amount of copper on the site suggested immediate access to copper sources. The collection included scraps of partially worked copper as well as 29 artifacts: fishhooks, beads, awls, knives, chisels, a punch or pressure flaker, fish gorges, and what Brose termed an "effigy claw" (1970:134). Wear marks on the artifacts suggested a range of functions relating to preparation of other tools, including woodworking, piercing, and fishing. The "effigy claw" exhibited the finest surface work and bore no use marks. Choices of manufacturing techniques (folding and hammering or hammering alone) related mostly to the size of the original lump of copper rather than to the kind of tool desired.

Overall the Summer Island copper was similar to that found on many small copper-working sites around the Lake Superior basin (Clark 1995; Janzen 1968; Wright 1963). The lithic tools and waste matched, to a certain degree, those of other sites over a wide time range. Brose remarked that unlike sites to the south, utilitarian objects of copper appeared more frequently in areas adjacent to source copper than in areas distant, where ornamental uses seemed to dominate the Initial Woodland metalworking technologies. He concluded, "It is difficult to see the Initial Woodland copper industry exhibited at Summer Island as anything other than a continuation of the Archaic copper industry" (Brose 1970:136), which is borne out by obvious continuities between Summer Island and places such as the Chautauqua Grounds site.

At the multicomponent Timid Mink site in Iron County, Michigan, Ottawa National Forest archaeologists discovered the remains of another Initial Woodland house (Hill 1995). The site occupied the southeast shore of Lake Ottawa, in a region rich in diverse resources. The site consisted of a dense midden or sheet of packed rich organic layers of soil in the vicinity of a prehistoric structure. This soil type resulted from human behaviors such as repeated trampling and compression, spilling of organic materials and wastes, and incorporation of food remains, broken objects, and decayed vegetation. Areas within the structure exhibited the most apparent midden development. The house was about 5 m long by 3 m wide, and was constructed of bent poles secured in the ground, lashed together, and then sheathed with bark sheets, on the order of historic Ojibwa houses of the region. It probably sheltered a single family. It was smaller than the Summer Island houses, and the posts composing its frame were also smaller, on the order of <10 cm in diameter.

The association of a concentration of quartz flaking waste and a copper awl or flaker duplicated the pattern of copper and lithic use from Summer Island, although the local rock type was different. This difference was expected, because prehistoric chert use gave way to quartz use as one moved west and north onto the Canadian Shield. Vast quantities of quartz debitage and expedient tools were recovered, as well as prehistoric ceramics typical of the Door Peninsula of Wisconsin. Radiocarbon assays of the Timid Mink midden indicated an age of ca. 1500–1700 BP. Food remains such as elderberry and raspberry seeds were also found at the site. The presence of these plants plus the location of the site exposed to onshore breezes of Lake Ottawa suggested a summertime use for the house by a single family of three or four persons. A composite map of the midden, associated artifacts, and structural remains showed that the midden encircled the house, at the center of which was the hearth (Hill 1995:355). Fire-cracked rock and other artifacts littered the hearth; an entryway path into the house was archaeologically visible because it was clear of clutter. Activities such as quartz working and pottery using were also apparent from the distribution of their remains within the structure. Some evidence suggested that the house may have been encircled by a small earthen berm on the exterior, a pattern well known for historic native people in the region (362). The association of copper awls or flakers and quartz repeated the suggestion that these tools functioned in a range of ways, including the shaping and/or sharpening of stone tools.

Initial Woodland Life in the Lake Superior Basin

On the south shore of Lake Superior there existed additional evidence of the continuity of Woodland life with patterns of the past, such as the inclination to use copper in everyday activities. The multicomponent Naomikong Point site sat adjacent to the prolific fishery of the shallows of Whitefish Bay. The site was partially excavated by the University of Michigan Museum of Anthropology in the late 1960s (Janzen 1968). Overall, the site was more frequently occupied than the Initial Woodland sites described earlier, which may be confirmation of the richness of the local fishery. Much of the pottery from the site was Laurel Ware, and a radiocarbon assay from the site suggested an occupation date of 1500 BP.

The lithic assemblage consisted of local cherts as well as several flakes of obsidian, perhaps obtained by trade, from origin points in Wyoming (57–58). Scrapers were the most common tool category. Hundreds of flat, notched stones were found that may have served as net weights. Patterns of posts near hearths suggested that the people of the sites smoked their fish over wooden racks. Bone remains could not be identified to the species level. But mammalian bones were more common than fish or bird remains. Despite the site setting at the lakeshore, this finding was expected, given the relative robusticity and size advantage that mammalian bones hold over fish bones in terms of resisting biodegradation. There were no floral remains reported from the site. It appeared to contain clusters of artifact types that indicated places where people manufactured chert artifacts such as scrapers and flake tools.

The copper assemblage consisted of beads, awls, and fishhooks. The presence of partially worked pieces of copper argued for the use of the site as a copper-working locality. Copper was not important at Naomikong to the extent that it was at Summer Island, despite their similar ages of occupation. Initial Woodland sites on the north shore of Lake Superior also exhibited the region-wide tendency toward frequent reoccupation and contained copper subassemblages of awls, gorges, beads, and chisels (Wright 1963).

Elsewhere, interest in the use of copper thrived. To the south of the Upper Great Lakes, spectacular decorative and artistic depictions were rendered in copper among Ohio Hopewell mortuary followers. Fitting (1979) connected the southern affection for copper with Initial Woodland traders who occupied northern Michigan's Straits region as early as 1700 BP. The best evidence available for Initial Woodland period copper working came from the site 20KE20, Keweenaw County, Michigan. The details of the discovery of an artifact cluster or cache of copper materials there have been reported elsewhere (S. Martin 1993, 1994). The cluster consisted of a group of 43 awls of various sizes, a group of strung beads (n = ca. 300), a crescent knife, and a large volume of partially worked copper fragments estimated to weigh about 13 kg. Many of these fragments fell into the size ranges of finished beads and presumably were preforms for bead manufacture. Partial oxidation of the copper materials offered a perfect microenvironment for the preservation of organics; within the deposit were found plant macrofossils and seeds. But the most interesting constituents were a woven textile fragment and a leather bag or packet within which all of the copper materials were originally contained. The leather yielded a radiocarbon assay of 1570 +/– 100 BP (uncal, B-24330). No species identification of the leather was possible.

The textile was of a technology and type widespread across prehistoric eastern North America. Such textiles are only found in archaeological contexts that enhance organic preservation, such as dry caves, inundated sites, and immediate contact with copper oxides. Therefore, a textile find from the Upper Great Lakes region is a rarity. The fibers of the textile derived from the inside bark of native plants and shrubs such as dogbane, milkweed, and basswood. Lengths of prepared fibers were spaced and suspended from a simple loom, generally a stick hung in a fixed position. These warp fibers were then wrapped by two weft pieces or horizontal interwoven fibers tied in a pattern or patterns alternating under and over the warp. The weft fibers were similar to common milkweed (Bisbing and S. Martin 1993:173). There were two or three additional constituents to the textile. One was a clear nonfiber material woven around the warp; it was translucent and appeared to be porcupine quill, a favored decoration used by the indigenous people of the Upper Great Lakes in historic times. There was also a pigment difference in the edge of the textile that looked as though some fibers were dyed as decoration. Similar textile remains were recovered at the Riverside site (King 1968).

The discovery at 20KE20 of a crescent knife within a cluster of artifacts dated to Initial Woodland times was unexpected. The conventional wisdom about such artifacts was that they belonged to the time period of the OCC, not to the more recent Initial Woodland Stage. Some other explanations were possible. One was that the crescent was an heirloom artifact from the OCC collected by a later person. Another was that estimates of the ages of crescents were not well founded and that they were artifact types that were younger (more recent) than the OCC. A final one was that the crescent, like the awl and the bead, was an artifact type that had a long life span in prehistory. We currently do not know which is the correct conclusion. Only excavation and dating in context will ever solve this question to the satisfaction of archaeologists.

The other lingering question at 20KE20 had to do with the identity of the copper packet's prehistoric owner. Its contents were all objects whose uses suggested sedentary activity such as fabricating/wearing of beads, scraping/cutting with a crescent, and piercing or chipping with an awl (Leader 1988). Such activities contrasted with artifacts whose assumed functions implied physical action: copper spear points, fishing gear, pikes, gouges. If we better understood the full range of functions for these implements, we would be in a better position to identify the potential prehistoric user of the artifact cluster. The 20KE20 packet may have been the property of a woman (S. Martin 1994). This interpretation rests on the assumption that awls and crescents were women's tools and that women were most likely the makers of fine beadwork. But awls and crescents were known to occur in male burials. Awls in particular probably served as do-everything tools as well as rough blanks for the production of other objects. Hill found awls in association with lithic chipping debris at the Timid Mink site, as did Brose at Summer Island. Archaeologists do not, at this point, fully understand the range of functions of many prehistoric tools. Whether this deposit belonged to a woman is uncertain.

Isle Royale bears evidence of Initial Woodland occupation, in the form of general Laurel Ware pottery. There appeared to be no significant adaptive differences between life ways of Archaic and Initial Woodland people, except that ceramic technology was unevenly adopted and chert tools replaced and/or complemented those of quartzite. On Isle Royale copper tools continued to be a part of the technological order of things. Clark (1995) found no apparent material links between Isle Royale's Initial Woodland people and the flamboyant use of copper in the lower Midwest's Hopewell Interaction Sphere. He suggested that adjacent groups of people may have acted as intermediaries in hand-to-hand copper trade, which might have served to isolate Lake Superior people from direct contact with southerners (167).

Robert Salzer found evidence of these intermediaries in his work at the North Lakes region of the Wisconsin/Michigan border (1974:49). In the sites of the Initial Woodland Nokomis Phase there, Salzer found "a heavy reliance upon trade for exotic ceramic and lithic raw materials" (49). The evidence for this claim lay in the ceramic and lithic subassemblages at Nokomis sites. They included pottery that was either imported or copied from pottery styles originating in the south and east of the Lower Great Lakes, as well as chert varieties that originated in Illinois. Waste flakes of obsidian were sometimes found in Nokomis sites. Copper working was widely evidenced. Hearths, fragmentary copper waste, anvils and hammerstones, a range of artifact types, and copper ingots or blanks were found in profusion.[8] The artifacts associated with Nokomis sites indicated continuity with other sites in the region: beads, awls, conical harpoons, fishhooks, beveled spear points, chisels, and punches.

Nokomis sites were larger and more intensely occupied than those of earlier times, which suggested to Salzer that there were more people in the region than before. The site locations were associated with inland lakeshores, lake outlets, islands, and riverbanks. At the Robinson site, excavators discovered an oval semisubterranean pit house ca. 3 x 4 m in size, supported by central posts. Refuse patterns helped to identify this feature as a house. Like the house at Timid Mink, it was built below grade. Similar houses were discovered in Door County, Wisconsin, at the Richter site (Stoltman 1986:274).

The most interesting question that remained about Initial Woodland life was the connection, if any, between the people of the copper-bearing areas of the Canadian Shield and the flashy consumers of copper who occupied the riverine lower Midwest. We know that copper appeared in quantity in the mortuary deposits of the Ohio Hopewell and related peoples. What we do not know is how they acquired it, or what they traded for it. It is not completely clear that the southerners used copper from the lake district at all. Some researchers preferred the idea that the glacial drift of Illinois, Indiana, and Ohio provided adequate sources for Hopewell demands for copper.

The Terminal Woodland Stage

By about 1,300 years ago, expanded intercultural contacts occurred in the Upper Great Lakes, which brought native agricultural plants into the area. The period from about 1300–400 BP was a time of relatively rapid change in the Upper Great Lakes and elsewhere, marked by subsistence intensification, population growth, and increased intergroup hostilities. In the Lake Superior region the period was different from earlier times, at least archaeologically, because more and more items originating outside of the region found their way into local archaeological deposits. This suggested that people of these times were interacting in novel ways, or more frequently, or both. Based on archaeological deposits at the Straits of Mackinac we know that agricultural products such as corn were becoming part of local food supplies. How this happened is not well understood. The northerly appearance of corn in prehistory suggested trade accompanied by a developing interest in horticulture and related food shortages (Lovis et al. 1996).

The best known site, and the one that literally defined what the local Terminal Woodland Stage was all about, derived from Bois Blanc Island in the Straits of Mackinac (fig. 24). At the Juntunen site prehistoric people began a long history of occupation as early as ca. 2000 BP. A model excavation from the 1960s, the site was meticulously excavated and interpreted by Alan McPherron, then a graduate student at the University of Michigan (McPherron 1967). Between ca. 1100–600 BP a series of as many as 25 separate occupations created a stratified sequence of archaeological remains at the site. One reason for the frequent use of the Juntunen site was a rather stable subsistence adaptation to the prolific fishery that existed in the nearby Straits of Mackinac. At virtually any season of the year one or another edible fish was able to be captured. People used all sorts of fishing gear and methods to capture the fish (Cleland 1982, 1989; S. Martin 1989). At least 13 species of fish were found among the Juntunen site remains, and fish bones, identifiable or otherwise, accounted for 85 percent of the bones recovered.

During its more than 600 years of intermittent occupation, pottery styles changed noticeably at the site, and these changes had a number of related causes. Some were due to the fact that people of more than one archaeological culture occupied the island. Others were the result not only of the changing imaginations of the potters, but also of the influences those potters may have absorbed from people of other social groups. Finally, some of the changes were probably due to trading of pots among people. Potsherds that were most frequently found in the Mississippi Valley, Ontario, New York, and Illinois were also found, in limited numbers at least, at Juntunen and other contemporary sites. This suggested a dynamic set of social groups whose lives were busy with interactions of many kinds: festivals, burials, marriages, trading events, conflicts. In contrast to the variability of the pots, the stone or lithic materials used at the Juntunen site were dominated by stone available locally, augmented by cherts gathered from the Thumb area of lower Michigan. The use of southern cherts suggested active interactions with people from Lower Michigan, or travels there to obtain cherts, or both.

At the time of its discovery, the remains of copper industry at the Juntunen site was one of the largest collections of prehistoric copper known from one site. The trait that distinguished this collection from those of earlier times, including the OCC, was that it was composed of diminutive tools and fragments. For this reason McPherron called it, somewhat tongue-in-cheek, the "New Copper Culture" (1967:164). The most recent (ca. 800–600 BP) occupations at Juntunen included more copper handiwork than earlier strata, but most copper could not be assigned to a particular occupation. Overall similarities between the pottery assemblages of the more recent strata at Juntunen and the pottery from Isle Royale partly convinced McPherron that the copper industry at Juntunen linked the people that lived there directly with Isle Royale. He concluded that Juntunen copper was related to an expansion of trade across the larger region (166, 258).

The technologies of working copper at Juntunen and elsewhere were essentially identical with those of earlier times: hammering and annealing. Many artifacts were produced by pounding the material into sheets and then folding them back upon themselves to form the final shape of the item. Some were made by using multiple sheets in the same manner and interlocking the sheets through repeated pounding. All stages of manufacture were represented at Juntunen. No specific copper-working tools were recovered. It appeared that hammers for copper working were not necessarily unique, but more likely indistinguishable from tools to work other materials, such as stone. The copper at Juntunen was concentrated in two small areas that McPherron interpreted as the interior of a longhouse close to hearths.

There were some data about structures at the Juntunen site that contrasted with the small houses recovered at Initial Woodland habitations. Juntunen's excavation revealed aligned series of postmolds, some containing the rotted remains of posts, which followed the contours of the site's land surface. Within these apparent enclosures were patterned dense clusters of artifacts: copper fragments, remains of hearths, flint/chert chips, trampled potsherds. Outside the lines of posts were very few artifacts. McPherron interpreted these patterns as large houses, including one that was at least 6.5–7 m wide. It probably contained multiple hearths and had been rebuilt or repaired on occasion. It looked like the house styles from prehistoric and historic Ontario Iroquois villages to the east of the Straits. The Juntunen longhouse was probably occupied about 700 years ago.

The excavation of human burials from the Juntunen site revealed information about life span, population, health status, and cultural connections with neighboring social groups. In contrast to earlier times, most of the burials at the Juntunen site were multiple secondary interments. In this burial style, thought to be widespread among hunter-gatherers of the Upper Great Lakes, human remains were bundled and curated by families, until such a time that they were collectively placed, probably with other kin and associates, in a common grave, or ossuary. At Juntunen, 65 persons were discovered in 7 ossuaries. At least one of the ossuaries was covered with an artificial mound of earth. The physical conditions of the people suggested something about lifestyle and normal physical activities. Women in particular appeared to suffer from bone maladies, which suggested that they spent much time using their hands in ways that may have encouraged soft tissue infection, perhaps in activities such as cutting, scraping, and perforating. Men, in contrast, were more likely to have suffered compression fractures caused by heavy work. Most people suffered from dental caries and atrophy. According to the numbers of individuals buried in comparison to population structures of living populations, the entire population of the island probably ranged between 25–50 persons.

McPherron linked the ossuaries with the historically known ceremonial, the "Feast of the Dead," chronicled by Brebeuf, Champlain, Cadillac, and others during the seventeenth century (290). Ossuary burial was also common during historic times among Ontario Iroquoian groups (229). The ceremony, and its attendant rite, that of co-burial, strengthened family, neighborly, and political/economic ties among people of different cultures and perhaps language groups. The ceremony apparently began long before the times of European observance. The longhouse housing style, the occurrence of eastern-influenced pottery, and the habit of ossuary burials led McPherron to believe that the most recent occupations of the Juntunen site demonstrated the connection of Upper Great Lakes peoples with Ontario Iroquoian neighbors to the east.

Elsewhere in the Lake Superior region, Clark (1995) found that by 800 years ago people of many regional social groups visited Isle Royale, presumably to camp, fish, and gather copper. He identified representative ceramics there similar to those of peoples from Manitoba to eastern Ontario. Individual sites included mixed assemblages of ceramics. Clark (171) suggested that the mixing was caused by trade as well as intergroup associations such as marriage. Such mixing was normal at Terminal Woodland sites on the north shore of Lake Superior, suggesting many years of contact between peoples from west and east.

Based on the number of Isle Royale sites and their contents, Clark concluded that small family groups used the island occasionally for food gathering as well as the quarrying of copper. Isle Royale's rugged shoreline restricted settlement to flat, gravelly beaches at the shore, and these places were frequently reused. Typical of the Upper Great Lakes, reused locations probably had relatively comfortable physical settings, diverse resources, and cultural traditions of prior use (S. Martin 1985). At Isle Royale, sites also sometimes occurred near copper mines. The data suggested spring-to-fall exploitation by small groups focusing on locally available foods such as beaver, fish species such as lake trout, and occasional large animals (Clark 1995). Though use of the island and inferred interest in copper appeared to peak in the Terminal Woodland Stage, there was evidence to imply that copper's use was mostly a local phenomenon. Clark surveyed published reports from archaeological research in the Lake Superior basin and portions of the Upper Great Lakes and discovered that the frequency of copper as a constituent of prehistoric assemblages gradually decreased in proportion to greater distances from the copper deposits. Overall, Clark suggested that copper use in the Terminal Woodland existed throughout the basin, but interest in it was culturally variable.

The Sand Point site (20BG14) was an 800–500 BP Lakes Phase village area and burial mound complex with a sizable copper tool and ornament industry and an apparent reliance on the fishery of Lake Superior's Keweenaw Bay (Holman and T. Martin 1980). In addition people relied on such plants as acorns, lambs' quarter, and a range of berries (20). Animal remains included small and medium-sized mammals as well as fall-spawning fish. Ceramic remains indicated ties with a wide mix of archaeological cultures. Western Michigan University excavated portions of mounds and habitation areas at the site in the early 1970s and recorded the remains of 117 people as well as documenting a "rich and diverse" material culture, including ceramic, lithic and copper technologies (15). The copper remains were generally scattered within midden and habitation debris as well as in the general fill of the site. In at least four cases, a tool (projectile point) and ornaments (crescent, beaded necklace, spiral ornament) or decorative copper items were associated with burials removed from the ossuaries at the top of Mound 1.

The copper industry at Sand Point was similar to that of the Juntunen site and included small tools, ornaments, and discarded fragments. The small sizes of the copper artifacts and fragments implied that the copper was casually collected as part of float or drift deposits, rather than mined or quarried in large segments. Technologically the industry was identical to those of earlier sites within the Lake Superior basin and showed no discontinuities of function or form. Both utilitarian tools (for wood working, leather working, and fishing) and ornaments were fabricated. The ornament forms echoed both longstanding tradition (a crescent and beads) and widespread stylistic similarities (the spiral ornament). The spiral was an ornament discovered in widely separate contexts: prehistoric Wisconsin villages (Salzer 1974), Terminal Woodland burials (Two Rivers, Manitowoc County, Wisconsin) prehistoric Minnesota mounds (Stoltman 1973), and contact-era eastern U.S. sites (Childs 1994).[9] The meanings represented by these forms were uncertain.

There were no clear indications of associations between the artifacts and a particular individual or individuals. Very few of the people interred at

Sand Point were buried with copper items. In contrast with other regional burial places such as the Riverside cemetery, very few children were found among the burials. The number of adult females in the 20–30 age range suggested that mortality during and following childbearing may have been high. The people suffered traumas such as fractures during life. There was also evidence of violence, in the form of an embedded projectile point in one male adult that had caused his death. Other people suffered from degenerative joint disease, bone inflammation, tooth abscesses, and periodontal disease.

Salzer (1986) described other Lakes Phase sites in northern Wisconsin. He suggested that by Terminal Woodland times, small, mobile, family-organized groups repeatedly used sites at interior lakeshores and stream outlets. These Lakes Phase sites included ceramics from a range of peoples, a fact Salzer connected with a trade relationship featuring Lakes Phase interests in copper (307). The copper industry of the interior, at least as represented by the Lakes Phase, consisted of small tools (awls, fishhooks, flat projectile points, tanged knives) and ornaments (spirals) produced by technological methods consistent with those found elsewhere in the region. Salzer suggested continuity between the Lakes Phase people and the historic Menominee, an identity earlier proposed by West (1929).

When/Why Did Native People Stop Using Copper?

Native copper was worked and used until the introduction of European metals in the seventeenth century (Fox et al. 1995), and though there were without a doubt different cultural emphases over time and space, there was no hiatus in the sequence of copper use over 6000+ years. The evidence from the basin's radiocarbon sequence confirmed that copper occurred at sites ranging in age from seven thousand years ago to the seventeenth century A.D. (Martin 1995). The radiocarbon sequence suggested that perceptions of copper's demise as a useful material were due to incomplete data rather than solely to culture change.

Research on copper artifacts from twenty sites in southern Ontario, reliably dated to fourteenth–seventeenth century contexts, demonstrated the active interest in copper working maintained by people of this region. When constant interaction with Europeans began in the sixteenth century, there occurred rapid replacement of native metals with materials from Europe (Fox et al. 1995). At the Dumaw Creek site (fig. 24) in lower Michigan and at the Pic River site in Ontario, Quimby recorded the co-occurrence of native copper implements and historic or protohistoric assemblages (Quimby 1966). These findings were also consistent with the protohistoric archaeological record from places such as the Gros Cap site, Michigan (20MK6).

At Gros Cap, traditional native design elements were expressed on novel materials such as glass and iron, acquired from Europeans (Martin 1979).

Concluding Remarks

This chapter began with an admonition about the indeterminacy of archaeological data, but despite that reality, it is apparent that carefully excavated archaeological sites tell lengthy, detailed stories about prehistoric life. Only meticulously excavated and carefully recorded archaeological remains really enable an understanding of such subtleties as changing house sizes and shapes, changing tastes in ornament design, sizes and age ranges of people's families, the foods prepared and consumed, health and life span, and sometimes the conditions under which people died.

Humans colonized the Upper Great Lakes region nearly ten thousand years ago. During this tenure they developed remarkably consistent means of making successful adaptations to local conditions. Their choices of habitation, the location of their cemeteries, their uses of resources, their impact on local microenvirons, and their social interactions reflected the long periods of stability and continuity of these adaptations. In fact there were no apparent disjuncts, disconformities, or anomalies in the history of the region's human occupations until the abrupt appearance of European cultures in the seventeenth century.

Despite the stability of local life, flexible and dynamic alterations took place over the many millennia of human experience in the region. The people were late partakers in moundbuilding, agriculture, and ceramic technologies, compared to their contemporaries in the south. Copper's importance as a technological, social, and ideological material waxed and waned, several times, over thousands of years. A great range of cultural groups occupied the Lake Superior basin and judging from the contents of their habitation sites, enjoyed one another's company or, at minimum, did not often get in each other's way. The repeated inclusion of exotic materials from neighboring regions in the archaeological assemblages of the basin's indigenous people suggests that external economic, political, and spiritual relations with people from neighboring regions were important events.

Suggestions for Further Reading

Halsey, John R., ed. *Retrieving Michigan's Buried Past.* Bloomfield Hills, Mich.: Cranbrook Insitute of Science, forthcoming. Summary of Michigan archaeology, with contributions by many of the state's professional archaeologists.

Holman, Margaret B., Janet G. Brashler, and Kathryn E. Parker. *Investigating the Archaeological Record of the Great Lakes State: Essays in Honor of Elizabeth Baldwin Garland.* Kalamazoo: New Issues Press, Western Michigan University, 1996. Eleven archaeology articles by twenty-one authors cover the Pleistocene to the recent past, as well as all regions of the state of Michigan.

Mason, Ronald J. *Great Lakes Archaeology.* New York: Academic Press, 1981. A look at the regional prehistory of the five Great Lakes, featuring archaeological research on time periods from early Paleoindian times to the historic era.

The Michigan Archaeologist. A quarterly publication of the Michigan Archaeological Society, this journal includes short articles of interest to followers of Upper Great Lakes archaeology. Occasionally, site reports are published in their entirety.

Ontario Archaeology. A twice-yearly publication of the Ontario Archaeological Society. Scholarly articles about Ontario sites, excavations and research projects.

The Wisconsin Archeologist. A quarterly publication of the Wisconsin Archeological Society, with coverage of adjacent states and provinces. Particular issues of interest include vol. 72, nos. 3, 4 (1991), a thematic issue covering the Paleoindian Stage in Wisconsin, and vol. 76, nos. 3, 4 (1995), whose theme was the archaeology of the Lake Superior basin.

Wright, James V. *A History of the Native Peoples of Canada, Volume 1: (10,000–1,000 B.C.).* Hull: Canadian Museum of Civilization, 1995. This compendium, based on primary archaeological evidence, covers all of Canada as well as important links with areas to the south. Two forthcoming volumes will extend coverage to the time of European contact.

7

Wonderful Power

Trading Copper

> The dwellers on Lake Superior seem to feel the most superstitious reverence for copper, which is so often found on the surface-soil in a remarkable state of purity. They frequently carry small pieces of copper ore about with them in their medicine-bag; they are carefully wrapped up in paper, handed down from father to son, and wonderful power is ascribed to them.
> —JOHANN KOHL

Introduction

The significance of copper depends upon the community and culture in which one lives. To some people, copper is an internationally valued commodity exchanged on mineral markets throughout the industrialized world: a capital-intensive industrial undertaking subject to boom-bust cycles that can make or break the fortunes of investors and workers alike. To others, copper exploitation is an environmentally destructive activity that destroys the quality of water and air as well as that of living communities. Still others believe in the essential therapeutic benefits of trace amounts of copper or the analgesic effects of copper bracelets against arthritis. In Michigan's Upper Peninsula, copper brings to mind the impact of early immigrant settlement and the short-lived rush to exploit the natural products of the region. Today, copper is, in part, a tourist commodity, in that the vestiges and lore of ancient mining and early industry attract visitors from many parts of the world to the Lake Superior copper district.

It is without question the case that copper's significance to the native people of the Lake Superior region was equally varied. The search for its meanings through oral histories, archaeological excavations, iconographic studies, and historical accounts yielded rich depictions of other worlds of causality, reality, and intrigue quite removed from the resource emphasis that nineteenth- and twentieth-century westerners placed on copper as a mineral. In protohistoric settings, copper contained unknowable quanta of power, which derived from its spiritual owners, the supernatural manitous. In order for humans to acquire and manage this power, intricate risky exchange arrangements combining trading skill, luck, and cleverness were made with these intangible spirit beings and, by extension, with other humans. The power conveyed by copper was a valuable commodity that could be transferred from person to person through trade in copper nuggets, charms, and finished artifacts. To some extent, the possession of copper items implied the possession and/or protection of the powerful but whimsical forces of the manitous. Evidenced by the geographical distribution of prehistoric copper objects, this power was sought by a myriad group of prehistoric communities, extending from the Upper Great Lakes to most of eastern North America. It was acquired and exchanged among trading partners across cultural, linguistic, and political boundaries for many millennia. Both otherworldly and worldly arts of the deal in copper are presented in this chapter, for they are really extensions of one another.

This chapter presents a review of what is known about the economic, social, and ideological significance of copper to the native people of the Lake Superior basin and beyond. The data to understand these meanings derive from archaeological excavation of objects and their contexts, the study of iconography (patterns of visual symbolism in artifacts, rock paintings, drawings, textiles, etc.), and the study of oral histories and written accounts of the native people of the region. The body of information about native beliefs and copper is most complete for recent times, the last four centuries. But comparisons of iconography from historic contexts with symbols from the prehistoric period suggest parallels in meaning. Accounts of seventeenth-century native beliefs and behaviors, combined with the results of archaeological research, suggest that traditions of reverence for copper are perhaps thousands of years old. These sentiments may have motivated trade and exchange of copper and other precious materials for many millennia. These speculative conclusions find support from a diverse data base whose patterning suggests conformities in the past over an expansive territory. The goal here is to link, through archaeology, the observed and recorded worlds of seventeenth- through twentieth-century belief with the cosmology of the more remote past.

In this chapter the reader will find a brief review of the geographical distribution of copper in what appear to be ritual contexts. The geographical area of concern is that environment of the Lake Superior basin and the Upper Great Lakes, the so-called Lake Forest region (Papworth 1967). The archaeological record provides a data base for a discussion of the commodities and accompanying mechanisms of trade that probably spread across the Upper Great Lakes. Then, we will consider the native ideologies about copper, which derive from the oral, historical, and material record of the Lake Superior basin.

Archaeological Research Regarding the Distribution of Copper

By at least four thousand years ago, copper implements and ornaments were widespread in prehistoric eastern North America. In some cases, the copper-using cultures also possessed exotic cherts, shells, and minerals. This distribution once puzzled archaeologists. They wondered how prehistoric communities acquired copper and other exotic materials in places so distant from their assumed origin points. One of the first comprehensive attempts to plot the spatial, temporal, and frequency distributions of prehistoric copper materials was completed by student geographer Ira L. Fogel in the early 1960s (Fogel 1963). Fogel addressed two very basic questions about copper: To what degree were there patterns of spatial and temporal difference in the use of copper among cultural groups? Did understanding these differences help to reassemble patterns of land use, trade, and culture change? Fogel faced many difficulties in his study, for he found it nearly impossible to establish firm archaeological context for many of the thousands of reported copper artifacts. They came from disturbed surface locations, from sites that had been destroyed, and from inadequately recorded localities. Associations among artifacts were vague or unavailable, and it was difficult to assess the age of the artifacts in the absence of any but inferential data. To deal with these data problems, Fogel's study looked only at artifacts from known archaeological contexts.

At the time, the prevailing view held that the archaeology of copper-using peoples could best be described by defining regional archaeological cultures, membership being determined on suites of artifacts and characteristics that were more or less similar. So-called diagnostic artifacts helped to pin down the identity of particular archaeological deposits and enabled them to be assigned to a cultural category. In this system, any site of Archaic age in which the use of copper was demonstrated, particularly in burial contexts, was assigned to one of several separate provisional archaeological cultures (fig. 25). The Old Copper Culture or Complex (OCC) consisted of a group of large, presumed utilitarian artifacts found most frequently in Wisconsin and nearby states and thought to be about three to five thousand years old (Ritzenthaler, "Old Copper" 1957). Some copper items were found in burials and appeared thereby to have ritual value to the peoples that used them. A somewhat similar category, the Red Ochre Culture (ROC) (Ritzenthaler and Quimby 1962) included burial sites that featured the liberal use of red

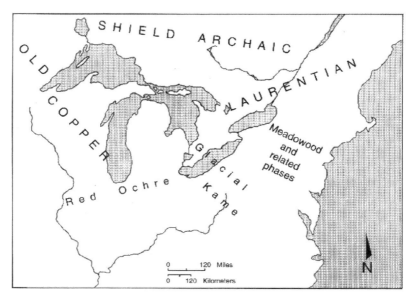

Fig. 25. Geographical distribution of some linked prehistoric mortuary cults using copper, as summarized in 1963. (Drawing by Brett A. Huntzinger, after Fogel 1963; used with permission from *Wisconsin Archeologist*.)

ochre pigment, the inclusion of exotic and distinctive chert blades as grave goods, the presence of copper ornaments and marine shell beads, occasional construction of mounds, inclusion of galena (lead), and a range of cremation and interment styles.

Another complex bore the title of the Glacial Kame Culture (GKC) (Cunningham 1948). Sites assigned to the GKC, very similar to those of the ROC, included a range of copper and shell ornaments, particularly a distinctive marine shell ornament known as a sandal-sole gorget. At the eastern extremes of the distribution of OCC, ROC, and GKC sites were other similar deposits, the Brewerton and Point Peninsula archaeological cultures. Researchers today agree that burial ceremonialism of the Middle and Late Archaic time period was widespread. Similarities in grave goods and apparent burial procedure suggested to many that ancient communities over a broad range of time and space shared and communicated a number of common beliefs.

Fogel tried to link certain artifact types with specific ecological settings within narrow time periods, but over a very broad area extending from the northern Midwest to New York State (1963:173). He found that there appeared to be statistically significant differences in the distribution of certain artifact types. Larger artifact types, such as projectile points and knives, were more closely associated with western sites, those near the assumed sources of raw copper. Ornaments were also associated with western sites. Small awls and gorgets typified the eastern and southern sites. He concluded that "functional correlation between regions and classes of artifacts are controlled by cultural and/or environmental factors" (158). Celts and adzes appeared in the east and in "the more distant reaches of copper distribution," which he related to differences in ecological setting and subsistence regimes (168).

Fogel confirmed a number of other spatial tendencies; the densities of copper artifacts in Wisconsin far outstripped those in other places. Those found on the surface were a poor reflection of those buried in primary context, which simply revealed that surface collectors were selective in what they chose to notice and pick up. There appeared to be a falloff in absolute frequency of artifacts the further one moved away from the Lake Superior source areas of copper, particularly for the discovery of surface artifacts. There also appeared to be an association between the distributions of copper artifacts and the major waterways of northeastern North America, a finding reported earlier by Popham and Emerson (1954:16) and later supported by Jury (1965, 1973). There were apparent differences in the degree to which people from east to west used copper in various times. There was evidence that both raw and finished copper moved through the broad area of copper's use. The mechanisms for this movement appeared to be trade routes of great age, which occurred across water routes connecting eastern North America with the Lake Superior sources of copper. By the end of the Late Archaic time period, cultural elaboration in the southern extremes of the distribution area, particularly in Illinois, appeared to eclipse the earlier importance of copper in the source areas and in Wisconsin. Fogel's overall conclusion was that "the distribution of copper in the Late Archaic period reflects an extremely complex pattern of cultural inter-relationships and environmental influences amongst and upon the cultural groups living in the Great Lakes region" (1963:173). In other words, it was not enough to categorize the varied archaeological deposits and thereby hope to explain them. Rather, it was important to focus on the complexity of human interactions in the region. Burial similarities suggested this complexity and interconnectivity; the distribution of the copper and other suspected ritual materials evidenced it as well.

Fortunately for progress in archaeology, some of Fogel's contemporaries had already begun the search for evidence of cultural interaction. The systematic study of copper began in regions adjacent to the source areas of the raw material (Popham and Emerson 1954). This work demonstrated that even outside of the presumed source areas, people apparently adapted copper and copper working to their own individual concepts, preferences, and idiosyncrasies. Ontario's local copper industry, likely derived from the technologies of Wisconsin and the Midwest, echoed familiar forms but of diminished size, suggesting experimentation with the incoming raw material. Some ornamental forms appeared to represent claw effigies; decorated copper bracelets were also recovered (12). There were three areas in Ontario in which there appeared to be concentrations of copper artifacts, which were interpreted as areas in which local cultures maintained a particular interest in copper. They were the Lake Nipigon region north of Lake Superior, the Thames River area of southwestern Ontario, and the Trent Waterway to the north of Lake Ontario. The latter areas coincided with historically known transportation and probable trade routes among neighboring peoples, which linked the Upper Great Lakes with the St. Lawrence River valley and regions to the east. Data supporting this model eventually came from Jury's research into the locations of copper caches in Ontario, which were spatially linked with routes of presumed trade (Jury 1965, 1973).

At the Morrison's Island-6 site near the border between Quebec and Ontario in the Ottawa River drainage, archaeologists found abundant evidence that copper working became part of local technology and activity by as early as 4700 BP (Kennedy 1966). This habitation and burial site was of interest because of its relative antiquity and because it lay far outside the known source regions for copper. Based on excavated materials, the people at the island lived by fishing and practiced complex technologies in bone, stone, antler, and copper. The copper implements included ornaments as well as items identified by probable function: eyed needles, projectiles, axes, fishhooks, gorges, punches, awls, and knives. Some of the copper, bone, and stone artifacts were decorated with parallel lines, dots, and zigzag lines, a common feature of copper and bone implements of comparable age in Wisconsin's burial contexts. The author concluded that though the site's materials were similar to many found locally in Ontario as well as south in New York State, the site's copper industry linked its people with their contemporaries in the copper source areas of the Lake Superior basin. Trade in raw copper and finished copper implements appeared to have taken place, in addition to the fabrication of locally designed implements such as needles, fishing tools and ornaments.

Fogel's analyses, Popham and Emerson's distribution study, and analyses of the Morrison's Island-6 and related sites suggested that the people of the vast region of the Upper Great Lakes and their neighbors shared a patterned set of behaviors and beliefs, which was evidenced in their cemeteries, featuring the occasional inclusion of special exotic artifacts acquired from one another through trade. This long-lived belief structure and its manifestations in burials varied through time and across the region, but indicated a coherence that must have come from constant interactions among communities in different environments and with different local commodities to offer one another.

Additional research on the archaeology of ritual and trade expanded the knowledge of the longevity of ancient communication networks and the interconnectedness of prehistoric communities across eastern North America (Baugh and Ericson 1994; Bourque 1981, 1994; Donaldson and Wortner 1995; Heckenberger, Petersen, and Basa, 1990; Heckenberger et al., 1990; Smith 1996; Varney and Pfeiffer 1995; Wright 1994). Copper's role in ritual uses and trading relationships was of great interest because of its durability and widespread distribution. In the following account of recent discoveries linking copper-using peoples, selected sites were included because they were well researched and updated Fogel's findings.

At the Hind site in southwestern Ontario, the excavation of thirty-seven individuals from a presumed GKC cemetery occurred a number of years ago (Donaldson and Wortner 1995; Varney and Pfeiffer 1995). The burials were interred between approximately 4500–1800 BP. As in the Wisconsin cemeteries discussed earlier, people were interred in a variety of ways, including primary burials, multiple burials, and cremations. Many burial inclusions (copper, galena, clay, shell beads and gorgets, utilitarian tools, animal bones, ground stone tools) were recovered as well. Some adult males appeared to be buried with copper axes, modified animal bones, and gorgets, although women and juveniles were buried with most of the copper beads. Men, women, and juveniles often were buried with shell beads of both local and marine derivation.

The people of the Hind site imported burial goods from distant areas; they also probably acquired some, such as galena, from near and far. The association of grave goods and burials suggested that people of both sexes and all ages owned exotic goods such as beads of shell. This suggested that they lived in a society without pronounced social hierarchies other than those produced by personal achievement. These conclusions drew on analyses of twenty other presumed GKC sites in Ontario. A systematic look at the area's combined record suggested variations on a theme of repeated use of fixed locales for burial over many years. These burials featured mixed styles of interment, an occasional association with copper and galena, the interment of perishables such as shell and animal bone, and the inclusion of red ochre and, later, pottery. The Hind site and associated sites were important because they underscored that indigenous Archaic mortuary ceremonies of great antiquity, developing in the northeast, probably lay the foundation for some of the burial ceremonialism of the ensuing Woodland era.

A look at the Boucher site and burial complex (Heckenberger, Petersen, and Basa, 1990; Heckenberger et al., 1990), located in northern Vermont, linked the behavior of its makers both to the mortuary complexes of the east coast (New Brunswick to Delaware) and to the midwestern source areas of copper. The Boucher site was contemporary with the Riverside site, which also, in common with midwestern cemeteries, saw repeated use over many centuries. This widespread pattern suggested to some researchers that burial areas were managed by territorial owners whose family groups used them over scores of years. Though the evidence for such use at Boucher was equivocal, researchers concluded that culturally varied local populations of people occupied traditional territories and marked them by burying their people at home, perhaps at more than one cemetery. This was similar to what researchers concluded about the Upper Great Lakes, at least at the Riverside cemetery (Thomas C. Pleger, personal communication, 1996). In this way the cemeteries were little different from other archaeological sites in the northeast and Upper Great Lakes, where persistent reuse through thousands of years was the local standard (Wright 1994:57).

In common with the cemeteries and complexes researched earlier, the Boucher site contained a highly variable set of interments. At Boucher there were between 84 and 100 people buried over a time range of nearly 2,000 years. Some of the variability in the style of burial interment was related to the long time spans represented at the burial sites, a situation that was also true of the Hind site and the OCC. At Boucher, single inhumation was the most common kind of burial, but cremations, particularly of adults, and multiple burials in groups of two or three were also part of the pattern. Sometimes cremations accompanied single interments. The most common grave inclusions were copper and shell beads. Fewer than 50 percent of the interments had beads, and their occurrence in cremations was infrequent compared with their presence in primary interments. Tiny copper beads were included in the burials of infants and children. The authors suggested that the source for the copper at Boucher was likely the Upper Great Lakes area. The apparent uniformity of manufacturing probably indicated "a single area of manufacture . . . likely at or near the source of the raw copper" (Heckenberger, Petersen, and Basa, 1990:192). The authors found no evidence that copper beads demonstrated a superior social position; rather, they suggested that beads proclaimed individual and group social identities (187). Shell beads were found more frequently with females than males, though uncertain determinations of sex made the association less powerful. The shell of which the beads were made represented at least four species whose origin points ranged from coastal North Carolina to Cape Cod. In some cases the beads were produced by careful drilling. Typically the shell beads were strung on fibers, and their appearance and numbers suggested local manufacture.

Copper artifacts other than beads were rare at the Boucher site. In general, artifacts other than beads appeared to be highly patterned in their associations with the interred people and apparently reflected age, sex, and differences in life activities. Adults were most commonly buried with what have been called, for lack of understanding perhaps, utilitarian objects. For example, a woven and hide bag stuffed with tiny copper scraps, which accompanied one burial, was interpreted as a craft specialist or ritual item (206). Though similar finds elsewhere were interpreted as craft specialist items such as bead blanks (S. Martin 1993), there may have been no distinction between craft specialty and ritual item in ancient belief systems. Likewise, animal bones could function as utilitarian food for the deceased, as well as ritual objects, insofar as they may have represented powerful animal spirit medicine. Some arguments for ritual meaning derived from the kind of animal interred; at Boucher, snake remains in pouches certainly suggested the latter interpretation (Heckenberger, Petersen, and Basa, 1990:197). At Boucher, some men were buried with such paraphernalia as well as shells and tools or tool kits; some women were associated with pipes and shell beads.

Despite its physical distance from the Upper Great Lakes and its long duration of use, there remained powerful similarities between Boucher and the Wisconsin sites with regard to themes of mortuary ceremony. Without a doubt there existed, across prehistoric northeastern North America, a continuous stream of human interaction, belief, behavior, and trade that is visible in archaeological deposits. Each local community possessed some level of boundedness, but at the same time was continuously linked with its neighbors in multiple directions. For this reason, assignment of the Boucher people to one or another archaeological culture was irrelevant. Heckenberger and colleagues concluded that "it is no longer fruitful to look for donor and recipient cultures. The task, as we view it, is not to determine if one thing (artifact, site, idea) is Middlesex or Meadowood, but to address the nature of the groups who were involved in the vast network of interregional communication which is clearly evidenced by the significant quantities of traded goods and pervasive similarities in mortuary ceremonialism" (Heckenberger et al. 1990:141).[1]

Some of the interregional communication occurred in trade. Hedican and McGlade (1993) found such widespread morphological similarities in so-called Old Copper projectile points that regional or subregional varieties simply could not be statistically defined. Homogeneity of form in the utilitarian sphere "lends support to the suggestion that the Archaic people in this area were closely linked by trade and possibly other forms of social contacts" (37). People of the Boucher territory probably gained copper through trade with midwestern cultures, but eastern copper sources, such as those in Nova Scotia, may have been used by some of their contemporaries. Geographic sourcing of Boucher copper proved inconclusive in one study (Wellman 1993); Levine's study of Boucher copper samples showed similarities with Lake Superior copper source profiles (1996:179). The people of Boucher obtained shells from coastal settings as far south as North Carolina. Some of the blocked-end tubular pipes at Boucher were made on nonlocal materials, as were ground stone objects discovered there. The presence of a walrus-tusk gorget with an obvious though unidentified origin elsewhere further evidenced widespread connectivity.

The motivations for participation in trade or the ideological constructs that rationalized trading behavior began at home, locally, rather than deriving from a monolithic and uniform set of beliefs. Yet the beliefs were compatible enough to enable and perhaps to require the existence of trade. Heckenberger's group suggested a chain as an appropriate analogy, consisting of independent links of down-the-line trading through reciprocal exchange, which carried on material, social, and ideological communication across the vastness of time and space. This system, though long-lived, likely underwent internal reorganizations of contrasting kinds at different times, changes that probably affected participating peoples differently, for instance people situated at opposite strands in the web of exchange. For example, during the third century A.D. at Michigan's 20KE20 site, someone prepared copper beads that were indistinguishable from those interred with Boucher children as much as a millennium earlier. By 300 A.D., the Boucher end of the trade continuum may have turned its attention to the south (Heckenberger et al. 1990:139). But at 20KE20, people may have traded with no notion that changes were afoot elsewhere. Their immediate trade partners were likely the same or similar to those of earlier years as was, to some degree or another, the product, but the copper ended up elsewhere down-the-line, perhaps in the Ohio Valley rather than in the Champlain lowland or the Trent Waterway.

Archaeological Evidence for Trade Near Copper Sources

Presumably, then, copper was a trade commodity from the Middle Archaic period onward and was found to the east, south, and west of the primary Lake Superior source areas in places and for durations of time that pointed to human agencies for its dispersal. Some, but hardly all, of this copper was known from probable ritual contexts or burials. Objects consisted of tools (both used and unused) and assumed ornaments. In some places recovery of raw copper suggested final fabrication of artifacts at localities distant from the source areas. Overall, variation in form suggested trade in raw material as well as finished copper items. It has been called "the most widely distributed nonperishable raw material in [the] prehistoric eastern United States" (Goad 1979:239). The evidence, however, for what exactly it was that came into the Lake Superior end of the complicated web of trade was not clear (Brose 1994).

Trade evidence can be difficult to establish with certainty. The best evidence for trade comes from raw materials that are found in clear archaeological contexts far removed from their likely points of origin in the natural world. The best items for determining relationships of trade are those that have unique or discrete sources and that are shown incontrovertibly to have come from one of a choice of specific geographical areas. Durable items are obviously more visible archaeologically than perishable ones, so artifacts made of metals, stones (lithics), ceramics, and minerals are of great interest (Wright 1994, 1995). Provided they survive from the past, artifacts of organic materials (shells, for example) can also help trace trade links, but the demands of pinpoint sourcing can be unenforceable beyond the general presumed home range of a species.

In the Upper Peninsula of Michigan, several sites provided a data base to investigate the trade importance of exotic items of lithics, minerals, and shell. Lithics were the most visible and most plentiful source of evidence for probable incoming trade goods. Fortunately many prehistoric stone sources were well known and localized as to origin. The western Upper Peninsula was a fortuitous place for the study of prehistoric trade in lithics, because its bedrock contained virtually none of the preferred kinds of lithics used by prehistoric people. The lithics trade was by no means unique to the region, however. By the Late Archaic time period, archaeological evidence documented busy exchange in lithics and other minerals up and down the rivers of the central part of the continent (Walthall et al. 1982).

Trade in lithics occurred very early in the history of human occupation of the Upper Peninsula (Brose 1994), as groups of gatherers and hunters moved through their seasonal collecting cycles, some of which involved acquisition and exchange of prized lithics such as Hixton orthoquartzite and jasper taconite (Clark 1991). The lithics of Isle Royale later on indicated acquisition of stone originating north and northwest of the island, but cherts originating east of the Lake Superior basin also underscored interconnnections in lithics use (23). This long-lived trade in lithics continued alongside the trade that intensified as Europeans came into the north. For example, catlinite (pipestone) from Minnesota became part of the lithic assemblage of Straits of Mackinac native sites by the seventeenth century (Cleland 1971) and perhaps even earlier at Riverside (Hruska 1967). Links with the northern Plains area trade, in obsidian, in chert, and finally in catlinite indicated that the east-west trade in lithics was a system of great longevity and structural stability into the historic period (Vehik and Baugh 1994). Not surprisingly, copper from the Upper Great Lakes was found there as early as the Middle Archaic time period.

There were numerous reports of obsidian from solid archaeological contexts in the Upper Peninsula of Michigan. An obsidian projectile point was collected from Isle Royale's Siskowit Lake by the McDonald-Massee expedition (West 1929). More recently, an unprovenienced projectile from the copper-bearing region was examined by Steven Shackley of the Phoebe Hearst Museum, University of California, Berkeley, who reported that its composition was typical of the Yellowstone deposits of Wyoming, but the elemental composition did not exactly match any known sources in the region (M. Steven Shackley, personal communication, 1996). Other obsidian materials included flakes from the Initial Woodland Naomikong Point and Winter sites (Janzen 1968; Richner 1973), a block from Riverside Cemetery (Hruska 1967; Papworth 1967), and waste flakes from Nokomis phase Middle Woodland sites in the North Lakes region (Salzer 1974). The Naomikong and Riverside specimens originated in formations known from Yellowstone (Griffin 1965). Prehistoric interest in obsidian, manifested by finds at Middle Woodland Hopewell sites in Illinois, Ohio, and elsewhere, is well known in the archaeological literature (see, for example, Baugh and Ericson 1994).

Hruska reported that six features at the Riverside Cemetery included what he identified as carefully flaked ceremonial blades of Indiana/Illinois hornstone. In some cases these blades were charred and cracked as a result of burning, "such as would be present in an intense cremation fire," though some may have been broken prior to burial (1967:176). Their orderly positioning evidenced intentional arrangement, and in one case, the position of the blades appeared to be the result of deposition within a bag or container. At Riverside these blades appeared in some of the most ornate burials. Most were thin, bipointed, and symmetrically leaf shaped, and their similarities suggested workmanship by a single individual (235). Hornstone was a prominent item in the burial ceremonialism of the ROC of the southern and western Great Lakes (Pleger 1996).

Twenty unworked galena cubes (lead sulphide ores) were discovered within the Osceola site fill. They were not identified by geological source nor were they associated with a particular burial or burials (Ritzenthaler,

"Osceola" 1957). Galena was available from Wisconsin sources in the Lake Koshkonong area, so its identity as a nonlocal material remained tentative. Galena was a favored trade item in the northern Plains, the Midwest, and the lower Mississippi (Brose 1994; Vehik and Baugh 1994; Walthall et al. 1982). It came from a number of sources, including Wisconsin, upstate New York, the headwaters of the Mississippi, and the southeast.

Archaeologists reported materials visually similar to white cherts of the Burlington formation (Iowa) from sites in Wisconsin and Michigan (Pleger 1996). Despite finding white cherts, Hruska did not identify this specific material at Riverside. But the dimensions of finished artifacts, the dearth of workable materials of such size in the local glacial drift, and the lack of debitage of such materials reported from sites in the region suggested that the artifacts were imported in finished form (Pleger 1996). At Riverside, Hruska identified a number of artifacts from the general fill and village area under the category "foreign or imported materials." These projectile points, scrapers, and blades were manufactured on Knife River flint, "which commonly occurs in North and South Dakota" (1967:247). The visual identification was not confirmed by thin-section or chemical analyses. Clark (1991:22) reported that materials formerly attributable to sources in the Dakotas were more likely native to the Hudson Bay lowlands. Genuine Knife River flint was widely traded along west-east routes to western Minnesota and east to southern Wisconsin (Vehik and Baugh 1994:258). Salzer (1974) reported it from Initial Woodland contexts in Wisconsin's North Lakes region.

Perishable trade commodities are routinely underrepresented in archaeological deposits, because their degradation removes them from the record. However, such things no doubt circulated within the trade networks of prehistoric North America (Baugh and Ericson 1994; Brose 1994; Smith 1996; Trigger 1987; G. A. Wright 1967; J. V. Wright 1995). Lacking archaeological data, written records help to infer what of organic origin was of interest in the trade. In the Upper Great Lakes, the written accounts of seventeenth-century French observers provided a partial restorative to otherwise conjectural information about native trade in perishables. An accounting from one area, that of northern Lake Huron, suggested the following perishable trade goods handled during the seventeenth century by the Odawa of Manitoulin Island: reed mats, fish, berries, meat, furs, and antler. In addition they acted as middlemen handlers or importers of nets, hemp, oils, tobacco, maize, wampum, marine shell, and medicine (Smith 1996). Ethnic groups to the west, such as the Menominee and Winnebago, handled bison skins (273), while the southern Huron were the primary source of maize, black squirrel, and raccoon skins (Trigger 1987:62–63). Trigger, Wright (1994), and Smith (1996) described in some detail the various trading arrangements that occupied seventeenth-century native people in the Upper Great Lakes. Together their accounts suggested that the trade links of native people at the time of the earliest written accounts extended across the eastern half of the continent.

Demonstrating the prehistoric presence of such commodities and explaining their roles in earlier trade requires painstaking excavation, recovery, and analytical methods. Sadly these methods were not always obvious, available, or carried out. In addition, it is unreasonable to project without any qualification the behavior of people in one century upon those of an earlier time and circumstance. Moreover, the presence of the French, related primarily to exploration and trade, was a factor in how native people perceived their own advantages and carried out their strategic alternatives. Thus what is learned from these sources provided hypothetical constructs rather than evidence about earlier trading patterns. However slim, these data help reconstruct what may have happened during prehistoric trade.

The primary archaeological evidence for trade in perishables prior to the seventeenth century in the Upper Great Lakes consisted of marine shell excavated from burial contexts. At the Riverside site, Hruska identified shell beads as "salt-water shell" but did not provide a species identification for them. Recent reanalysis confirmed that they were marine in origin (Thomas C. Pleger, personal communication, 1997). A range of species of Atlantic coast shells was also found in the burial suites at the Oconto and Reigh sites in Wisconsin. The evidence for the importance of the shell trade was stronger in the central Midwest, where it was associated with copper in burial settings (Goad 1979; Winters 1968). Winters noticed that over time there were long-wave fluctuations in the activity level of the shell trade in the central Midwest. The prolific acquisition of copper was clearly part of trade cycles there, where marine shell and copper were tightly associated. "Marine shell and copper were being distributed over vast distances within eastern North America by the third millennium B.C." (Winters 1968:219). Studies of the archaeological distribution of marine shell may be critical to understanding how prehistoric trade networks operated, particularly among the large communities of the American southeast during the late years of prehistory (Goodman 1984). The northern data for the shell trade are sparse. The acidic soils of the north ruled out the physical preservation of shells unless they were associated with very favorable microconditions such as the chemistry of copper oxidation (Janaway 1985).

Summary of Archaeological Evidence for Trade

Despite the slim yield of preserved and excavated prehistoric trade items from the archaeological record near source areas of copper, there is still a need for explanation when *any* recovered exotics demonstrate contact among distant peoples and places. Going beyond tangible trade goods, some researchers suggested that trade took place in commodities that were not material to begin with. Trade may also involve the exchange or purchase of songs, poems, social standing, magic, or ideas. Any understanding of trade requires some inferential arguments, some additions of analogies from observed groups in historic times, some recourse to documents, and some uncertainty.

Generalities about trade are clear from a historical comparative perspective. Virtually everywhere in the world, trading activity ensures logistical security, neighborly goodwill, and mutual protection. In some environments trade may have been an ecological necessity in order for people to occupy marginally productive territories or to reduce the risk of food shortages (McHale Milner 1994; Smith 1996). Environmental boundaries encouraged trade when desired objects or preferred commodities were unavailable locally but lay close by in the hands of a neighboring group. In addition to satisfying logistical issues in the short run and satiating the desire for unusual or unique objects, trade enhanced the satisfaction of a number of related goals. It enabled information exchange, and it could involve the search for a spouse, the spread of innovative ideas and technologies, the promotion of alliances, and the acquisition or bequest of important and powerful gifts. All kinds of commodities were given and acquired. Food, tools, decorations, and trifles may be visible evidence of trade. Medicines, charms, ideas, songs, and privileged or sacred information may not be so obvious. The particular cultural importance attached to a traded commodity was also not directly observable and may have differed between provider and recipient. Values were, after all, culturally derived, and ideas about what to trade and what to trade for were not necessarily uniform, even between individuals.

Nor were the mechanisms of trading uniform over time and across geographic space. The possibilities ranged from occasional face-to-face equitable exchanges between friends to scheduled ceremonies in which large quantities of precious goods were redistributed and/or consumed on the spot. Trading partnerships were forged and then continued through a range of institutions such as adoption, or the establishment of foster kin ties, or through the exchange of persons in marriage. Trade was sometimes guaranteed or prolonged through the mutual exchange of guests or hostages. Sometimes the terms of trade and one's actual access to it depended upon the commodity concerned or the social stratum to which one belonged. In some cases an individual family or kin group controlled the rights to a specific commodity or the specific route that the trade traditionally followed. Some groups controlled trade at strategic places. In extreme cases this control meant the development of a trade monopoly and the restriction of all but approved access into the source territories of a prized commodity. To complicate matters, sometimes multiple kinds and patterns of trade took place at the same time and place.

Presumably, differing mechanisms of trade leave discernible and contrasting patterns behind. Archaeologists sometimes study the variable patterns of trade and condense them into contrasting models of commodity distribution (Renfrew 1972). For example, Fogel measured the frequency of occurrence of Late Archaic copper artifacts over straight-line distances from two presumed source areas. He discovered that there was a rapid reduction in their frequencies at distances greater than 300 km from the presumed centers of acquisition and again at distances of greater than 550 km. The spatial patterning of these two falloff zones suggested to Fogel that the "centers of production and utilization occur at some distance from the source of the raw material" (1963:170). He was unable to confirm what sort of mechanism (trade, expeditionary trips to the source areas, etc.) best accounted for the dispersal patterns.

The double falloff pattern demonstrated by Fogel approximated what Renfrew (1972) called a *directional trading model*. In this model, items of trade were preferentially acquired by people at a distance from the sources, while people closer to the sources had little or none of the commodity. The distant people, in this model, were assumed to concentrate the commodity and then control its later distribution. There is some appeal to this apparent good fit with the Upper Great Lakes Late Archaic case, yet with qualification it looks less convincing. Excavated sites, the source of Fogel's data, were somewhat rare at the time of his work. Comparing surface-collected with excavated collections, Fogel discovered that the surface artifacts gave a radically different picture of copper distribution. There appeared to be no falloff until the 550 km interval. This single falloff pattern approximates a *down-the-line exchange* model. According to Renfrew's explanation, this pattern resulted when many local episodes of acquisition, use, and exchange made the commodity generally available. These many minor acts of exchange did not totally cease outside of the source area, but they became less frequent, and the archaeological record reflected this decrease in activity as a reduction in the amount of the commodity at any one place. This down-the-line model appears to fit the distributions of copper artifacts in the Midwest during the Late Archaic time period. It also makes sense given the actual geographical distribution of raw copper materials, which are widely available in the glacial drift deposits of the Lake Superior basin and southward.

Most researchers agree that at least during Late Archaic times, trade in Lake Superior copper likely occurred as "geographically-restricted down-the-line exchange" perhaps propelled by "small family group acquisition" (Brose 1994:219). In fact this model fits the distribution of copper until late in the Woodland era. More recent times reflected a shift both in how copper was used and in its likely value to those outside of the source areas (Brose 1994), as well as potential shifts in source areas themselves (Goad 1978). Tracked over half a continent and several millennia, the patterns of copper's distribution show these conditions of change.[2] Closer to the northern source areas however, there was a very gradual increase in the local population and a steady interest in acquiring copper. The evidence that trade mechanisms changed in operation is lacking, at least until the twelfth to fourteenth centuries A.D. There is, for example, no evidence of an external presence actively acquiring Lake Superior's resources until (1) the apparent spread of people influenced by Ontario Iroquoian groups and perhaps other horticulturalists into the basin about A.D. 1200–1400 and (2) the entrada of Europeans into the central part of the continent a few centuries later.

Ritual Beliefs and the Role of Prehistoric Trade

The long-lived prehistoric interest in copper and its occurrences in what appeared to be ritual contexts suggested that copper was a special substance that many people intentionally acquired. The urge to possess copper was linked with cosmological beliefs about sources of power. A look at Lake Superior basin mythology as recorded by the native people of the seventeenth–twentieth centuries reveals some of the spiritual significance attributed to copper.[3] This belief structure may have been part of prehistoric life as well, for all indications suggest long-lived continuity in the cultures of the region.

All people, everywhere, charge their physical surroundings with supernatural authority and, to some degree, predictability. This predictability has an internal orderliness or logic, as it is the cultural story of how things came to be as they really are. Obviously the nature of this reality varies from culture to culture. The kinds and characters of the supernatural authorities that ordain the cosmic order and maintain it also vary. For the native people of the Lake Superior region, reality was believed to be made up of three superimposed orderly layers or domains. Human life took place on an island, the middle layer, that was conceived to be floating within opposing domains of sky and water (Phillips 1984). The cosmos was also inhabited by powerful but conflicting entity-spirits or essences called manitous, who could move from layer to layer at will. Manitous influenced many of the events that people experienced in life, calamities as well as adventures and windfalls. Manitous operated within specialized spheres of significance and were different from one another in terms of their typical behaviors, likes, dislikes, powers, and appearances. Some were combative by nature while others were less so, but all were changeable and considered unpredictable. They could, in part, be cajoled, pleased, and manipulated by people, often by the presentation of gifts, offerings, and sacrifices. In effect these dealings were yet another trade relationship, in which people offered the manitous desired things in exchange for their beneficence or kindnesses. Vecsey characterized these dealings as taking place in a "very personal universe" (1983:73), in which men and women sought out the manitous for inspiration and power (Johnston 1995).

The behaviors and motivations of manitous were superficially similar to those of people; they enjoyed getting the best of one another through feats of strength as well as feats of trickery. They experienced greed, jealousies, lusts, vengeful desires, and selfish episodes and acted upon them at will. Their behaviors were capricious, and there were many ways of dealing with them. Encounters with manitous were situational, risky, and necessary for the acquisition of power and good fortune.

Manitous sometimes could be found sharing routine activities with people within the normal abode of humans. They enjoyed transforming themselves into ordinary objects or animals to disguise themselves from one another or from humans. As a result, various traces and incarnations of manitous were encountered throughout the middle world; rock formations, masses of copper, tree stumps, islands, bird calls and strange sounds, thunder and lightning, mists, waterfalls, storms and whirlpools were some of their many manifestations. One just never knew about the manitous (74). They were sometimes heroes, sometimes tricky, sometimes foolish, and sometimes deceitful. But they were always powerful, always present, and treated with deference, reverence, and gifts (xxi). One could not reasonably undertake any action without considering their impact on its outcome.

The essential behavior of manitous revolved around a series of eternal mythic struggles among adversaries: the sky manitous versus the underwater manitous. The sky manitous usually prevailed, but the water manitous offered desired powers as well. According to Warren, the evil spirits of the lakes and underground sometimes conspired to threaten the people of the region, but the vanquishing of dangerous lake fogs by powerful winds asserted that the sky manitous retained the upper hand (1984:93). Copper belonged to certain underwater manitous and it was they who dispensed its power.

Manitous created perilous conditions in their natural abodes. These threatening entities made some settings, such as lakes and islands, places to avoid or at least to be careful of. There are many reports, throughout the historical literature, of mythical islands and lakes, as these were believed to be the abodes of the evil manitous (Bourgeois 1994:42–45). The lakes themselves were perceived as abodes of powerful spirits who could be propitiated through proper sacrifices such as tobacco, white animals, red cloth, and fat dogs. Water features such as "perilous places in rivers and rapids" were also offered sacrifices (Kellogg 1917:112). The lakes and the islands within them were seen to behave in ways that were perfectly logical in local native cosmology. For example, they seemed to change shape and position as a result of mirage phenomena. Transformation was a widespread characteristic of Lake Superior native cosmology and was attributable to manitous; in fact it was one of their main traits. Everything transformed: water, copper, people, animals, islands, vegetation, manitous. Nothing was as it seemed, and it was imprudent to assume that the face of anything was its totality. Such themes have been analyzed in depth by art historians (Phillips 1984; Trevelyan 1987).

Though the manitou characters and their traits varied somewhat across the area under study, there existed an underlying similarity to versions of traditional belief about the powers of the cosmos. The general forms of these stories were widely dispersed over the Upper Great Lakes region. The following descriptions came from the people of the Lake Superior basin and southward during the seventeenth–twentieth centuries. They were collected by a wide range of observers and writers including Fr. Allouez (Kellogg 1917), Johann Kohl (1956), William Warren (1984), Homer Kidder (Bourgeois 1994), Victor Barnouw (1977), Christopher Vecsey (1983), Ruth Landes (1968), and Basil Johnston (1995).[4] These stories demonstrate that the presence of copper in the natural world was wholly integrated within the structure of native beliefs.

The manitous were more or less recognizable, named identities who specialized in certain realms and activities. Those of particular interest include Nanabozho (also known as the White Rabbit or Manabozho). Identified as the brother of birds (Bourgeois 1994:30), Nanabozho was closely linked with the world and power of the sky manitous, but was somewhat constrained by his earthly origin. He was the offspring of a human woman, Winonah, and a manitou; his father was the West Wind. Nanabozho was a hero who acted upon very human motives in sometimes disastrous ways. He was part bungler and part joker, especially when his human side ruled his actions. He possessed great powers of transformation and often turned himself into a pine tree, or a stump, or a white rabbit. The bird Shingebiss, the pied-billed grebe, was his ally, for every spring Shingebiss returned to the north carrying gifts of precious migis (beads or shells) acquired in the south (36). Nanabozho's enemies included Mishi Bizi and the other underworld creatures, because they were responsible for the death of Nanabozho's nephew, the Wolf.

The Mishi Bizi manitou was also known as the White Bear, the White Lynx, the Great Lynx, and the Underwater Panther. He was sometimes described or depicted as a snow white panther, but his home was the water. Mishi Bizi was believed to be the chief of all the water spirits, most of whom were considered threatening or downright dangerous. White animals, in general, were identified as his allies (Bourgeois 1994:19); frogs were his relatives and guards, and horned snakes were his companions. Mishi Bizi controlled the abundance of animals and had hegemony over objects within the earth. He was logically identified as the owner of copper. Allouez, writing in 1665, noted that the Ottawa "hold in very special veneration a certain fabulous animal which they have never seen, except in dreams, and which they call Missibizi, acknowledging it to be a great genius, and offering it sacrifices to obtain good sturgeon fishing" (Kellogg 1917:112). To the Menominee, the guardian of copper was an earth-dwelling white bear, with silvery hair and a copper tail (Brown 1939:39).

Mishi Ginabig (Great Serpent, Horned Snake) also abided in the underworld and underwater realm. He was believed to be the largest and strongest of the underwater manitous; Vecsey regarded him as interchangeable with Mishi Bizi (Vecsey 1983:74). Conway reported that in general Mishi Ginabig was synonymous with the general wicked devil, the Mutchimanitou (Conway and Conway 1990:47). Allouez claimed that the native people professed that "the evil spirit is in adders, dragons, and other monsters" (Kellogg 1917:112). Mishi Ginabig's physical form was terrible to behold; his scales, horns, and long, coiled tail were made of copper. His horns or antlers were described as big as those of a moose (Bourgeois 1994:43). Though these copper extremities were seen as one of Mishi Ginabig's most sinister features, they possessed great power and securing one of them meant that the power was one's own, to some extent (Barnouw 1977:133). Mishi Ginabig's tail was so powerful that it turned over canoes, created storms and whirlpools, and caused floods. He was responsible for drownings, falling through thin ice, and general harm, but could be entreated with appropriate ad hoc sacrifices in exchange for good hunting and fishing, safe passage, and medicinal powers to avert disease (133; Bourgeois 1994:42). Therefore goods for him were left at rock formations around Lake Superior known to be his dens or those of his allies (Conway and Conway 1990:49), in the hope of securing safe passage on the lake. Such entreating was necessary for travel to Isle Royale, as recalled by John Linklater: "He stated that his wife's grandmother, and his own grandfather, remembered coming to Isle Royale. The latter recalled the gathering on the Canadian shore and the ceremonies, the dance and the appeal to the Spirits, that were deemed necessary before the trip could be made" (Fox

1929:317–19). Though water beasts were often associated with harm to children, at least one fortunate girl was safely transported across a lake by a swimming Mishi Ginabig (Bourgeois 1994:52). Barnouw reported that serpents with horns colored the water of lakes, expected sacrifices of babies, and transferred their power to bad people (1977:135). Weapons of cedar and copper were sometimes particularly effective against them (132–33; 162–63).

The manitou owners of copper were so powerful that their actions determined the lay of the land. Certain copper-bearing rocks in Lake Superior, buried in sand during storms, were seen as clear evidence that manitous had again changed shapes, or so it was explained to the Jesuit Allouez (Kellogg 1917:105). A trip in 1669 to the magical island Michipicoten in the eastern Lake Superior basin evoked a warning issued to Fr. Dablon about the dangers there. The island was the home of a manitou who, provoked by careless thievery of his copper, was responsible for the deaths of a number of island visitors. This island also had a sinister reputation for growing lynxes and rabbits as large as dogs (Foster and Whitney 1850:10–11). Of course these were not ordinary animals, but the embodiments of the powerful Mishi Bizi and his nemesis, Nanabozho.

The upper world or sky was occupied and controlled by powerful creatures represented by Thunderbirds, or Animikig. These manitous were considered extremely powerful and desirable spirit supporters; their voices and wings were the sources of lightning and thunder. They were believed to be protectors of fortunate people and to dispense success in hunting and war. In Bourgeois's account, ancient people who left offerings to such creatures gained medicine from powerful roots that grew near their nests (1994:40). These Animikig were the cosmic adversaries of the creatures of the underground and underwater realm. They were generally successful in hunting and destroying their enemies, the water manitous, who were particularly vulnerable when they came to the surface of the waters during fair weather (42; Conway and Conway 1990:39).

In addition to these powerful creatures, a range of lesser beings inhabited rocky places, woods, rustic places, and water bodies. Johnston (1995) collectively referred to these spirits as manitoussiwuk. Some of these lesser creatures were associated with the presence of copper. Others were harmless miniature creatures; still others were depicted as mermaids or mermen. A special group of manitoussiwuk caused wakefulness at night and brought warnings or protected children. Johnston identified little copper manitous as mizauwabeekummoowuk (1995:151), who were mountain-dwellers responsible for come-uppances to haughty humans. Brown identified the copper mines as the home of an unnamed magic woman who, upon becoming angry with humans after some transgression, sank into the ground and took her precious copper with her (Brown 1939:38).

There is little doubt that copper was spiritually precious and powerful according to the beliefs of the Lake Superior basin's native people. Writing in 1665, Fr. Allouez reported that the native people kept copper pieces "as so many divinities, or as presents which the gods dwelling beneath the water have given them and on which their welfare is to depend" (Kellogg 1917:105). Allouez' informants later "say also that the little nuggets of copper which they find at the bottom of the water in the lake, or in the rivers emptying into it, are the riches of the gods who dwell in the depths of the earth" (113). These beliefs were persistent over the centuries. Much later, Brown reported that copper pebbles were believed to be the feces of the Thunderer manitous, or their eggs (Brown 1939:40). Others identified copper as the cradle boards and toys of the manitou's children (Foster and Whitney 1850:10–11) or the bones or scales of the manitou (Phillips 1986). Brown reported the belief that Nanabozho was a smith who forged toys of copper for his children at his home near Chequamegon (1939:40). Copper possessed the power to avert disease (Kohl 1956:424–25; Warren 1984:98–99), and masses of it were considered particularly holy and powerful.

These examples suggest that northern native beliefs logically connected the behaviors of manitous, the excesses of weather, the transformations of nature, and the exchange of precious objects with harmonious relations between people and manitous. In this worldview, copper was a sought-after commodity with the power to bring good fortune. The manitous were the sources of this power and were compensated for it.

Material Culture and Ritual Beliefs

Everywhere, and in every culture, people mark objects with meaningful designs, patterns, and slogans. This aspect of human behavior is ancient and universal and is regularly connected with ritual events and performances as well as with everyday occurrences. The patterns of such depictions, or their iconographies, are systematic, stylized, and recognizable codes that tell meaningful stories to the knowledgeable. Some researchers suggested that the icon systems of northern peoples were not only of ancient origin, but were persistently observable in material culture (Fitzgerald and Ramlukan 1995; Garrad 1995; Hall 1983; Hamell 1987; Phillips 1984, 1986; Rajnovich 1994; Trevelyan 1987). By reviewing a selection of the many occurrences of iconography on objects of material culture, both those recovered archaeologically and those known from the historic period, one may be able to understand more completely the systematic meanings of objects in cultural context. Ultimately this may lead to a more complete understanding of regional prehistoric cultures. In her studies of nineteenth-century Great Lakes area iconography, Phillips commented on what she termed "the consistent approach to design and a pervasive iconographic system" portrayed in many media (1984:30). These media, according to Fitzgerald and Ramlukan (1995), included design depictions on rock faces, birch bark, stone discs, bone objects, ceramics, paper, and wood. Copper (and later brass) belongs on this list as well (Bradley and Childs 1991; Brose 1970; Fox 1992; Quimby 1966; Reid 1995; Winn 1942). Though much regional iconographic material was without a doubt of eighteenth- and nineteenth-century age, those recovered archaeologically represented a broader time range. For example, painted depictions (so-called rock art) of cosmological themes indigenous to the Lake Superior basin may be 2,000 or more years old (Rajnovich 1994). A review of some of these depictions on varied media follows, derived from ethnographic accounts as well as the study of objects.

Textiles were often decorated with mythic themes and multiple design elements representing opposing cosmic forces. One measure of the ambiguity of powerful and changeable relationships with spirits and their many aliases and manifestations was the regularity with which items of native material culture came encoded with such motifs. A skin bag, club, or personal item of dress would be composed of stylized depictions of the two opposing powers, for example a twined bag with a panther on one side and a thunderbird on the other. This paired appearance underscored the double-edged nature of power potentially available to supplicants (Phillips 1984, 1986). Likewise clothing designs were seen as both an announcement of power and a protection against it (Phillips 1986:29).

Some designs were interpreted as shorthand motifs, such as representations of stylized power emanations or movements associated with underwater creatures (Phillips 1984). By the nineteenth century such motifs (spirals, concentric squares or circles, parallel wavy or castellated lines, and linked chevrons/serpentine motifs) were specifically associated with the underwater manitous at least in textile media such as woven and skin medicine bags. In contrast, zigzag designs and hourglass figures were associated with thunderers. Iconographic shorthand allowed individual makers of objects to portray elements of private visions and dreams on objects of personal and ritual significance (27).

Iconographic objects were also reported from the historic literature. Kidder's informants told him that sticks engraved with notches were tallies of beaver who could be taken if the powers of the Mishi Ginabig were invoked. According to Kidder's friends, trade with the manitou was initiated by the offering of a pipe of tobacco and a red cloth belt, which were laid upon the water to generate the interest of the spirit. The manitou appeared and the belt began to float and align itself to the supplicants. "A stick that Black Bass was holding in his hand showed a great many cuts, the meaning of which was that a great many beaver would be trapped. In the fall Black Bass and his friend went to Three Lakes and trapped all the beaver they wanted, but they did not take as many beaver as the number of cuts on the stick because their luck was a gift of the spirit" (Bourgeois 1994:44). Landes reported that decorated sticks, pieces of cloth, and tobacco were appropriate offerings to waters that were thawing or starting to freeze: in other words, waters that were transforming (1968:31–32).

Rock paintings exhibited comparable iconography. Rajnovich's recent account of Lake Superior rock art linked signs, language, and ideology into one widespread system of belief and behavior that united a vast territory over long reaches of time. This iconography, typically done in red ochre on rock faces, represented the seeking of education, information, and communication without which the power of the manitous was unreachable (1994:11). Within this complicated system of meaning, colors and materials, signs and symbols, the geographical location of ritual events and paintings, and sought-after knowledge or power were integrated and mutually represented, often with layers of simultaneous meaning. For example, a lattice icon represented both a spiritual bridge between cosmic layers and a level or degree of sacred society membership. Quartz veins within a rock face and incorporated into a picture referred to the power of stone as well as to mythic snakes. Often the powerful manitous, Mishi Bizi and Mishi Ginabig, were represented in their characteristic threatening roles. According to Rajnovich's research, Mishi Bizi and his cohort were particularly accessible at places where two material realms or levels of reality met, for example where rock faces met water (35). The pictures of Mishi Bizi were found at cliffs, or in caves, and the manitou was logically associated with blindness (103).

In rock depictions, Mishi Bizi and the snake or serpent Mishi Ginabig appeared with spiral tails, horns, and power lines, particularly parallel lines on the bodies and tails of the snake forms, and rib lines on the bodies of lynx forms. In general spirit lines across the body referred to the manitou (157). Rajnovich linked these drawings with dreams, hunting medicine, and the acquisition of power. They were also linked with a spirit power that resided within stone, which was believed to be accessible in flint and other lithic material. The connection between lithic power and hunting success is self-evident; Rajnovich linked them through the iconography of triangle shapes, which represented wolf's teeth as well as arrows (150–52). This link was also made more complex through the mythic association of the Wolf nephew of Nanabozho with the power of flint and the eventual ritual vanquishing of Mishi Bizi. Signs, stories, and materials were, in this system, simply different aspects of one another. Each operated as an overlay of meanings on the others.

Stone discs and associated items, recovered archaeologically, provided a measure of time depth for regional iconography. A number of researchers recovered items of iconography from archaeological sites in the general area known to have been occupied, in the historic period and into the prehistoric past, by related groups of native people who were traditional residents of long-lived standing. Two areas were particularly of interest for their revelations about symbolic depictions of manitous on stone discs (Cleland et al. 1984; Fitzgerald and Ramlukan 1995). Cleland reported that at four sites in the Thunder Bay region of northern lower Michigan, shale discs were recovered from multicomponent prehistoric sites. These discs, some of which were drilled to suspend as pendants, were likely associated with offerings for safe passage across the open waters of Lake Huron and protection against evil events. Cleland's group concluded that their "restricted appearance suggests short-lived special function" (1984:235) related to a ritual connection with the waters of the adjacent bay. Those discs that were inscribed showed figures that were identified as Mishi Bizi, as well as figures of thunderers, otters, beavers, moose, and trees. Sometimes a disc included multiple figures. Cleland et al. linked the distinctive iconography of the discs with native styles seen in rock art and religious scrolls, but also suggested that the local variants of the iconography were to be expected given their selectivity of function. Given the prehistoric artifacts with which they were associated, the icon systems extended at least as far back in time as the twelfth century.

Across Lake Huron, at the Hunter's Point site on the Bruce Peninsula of Ontario, Fitzgerald and Ramlukan (1995) reported the discovery of slate and shell objects bearing comparable symbols. As seen at the sites across the lake, the Hunter's Point site represented a range of past human uses. A slate pendant had iconography on both faces; one was interpreted as a thunderer and the other as Mishi Bizi, with horns and associated power lines radiating from his spiked back and long tail. A shell engraving represented a thunderer being. The objects were associated with both ritual items of historic times and objects as ancient as the Initial Woodland period. The appearance of these objects in places that were used over long reaches of time supported an interpretation of long-lived stability in symbolic systems. Fitzgerald and Ramlukan reported that a historic-period brass bracelet was recovered from the same site, which bore a depiction of a horned snake/underwater serpent. Garrad suggested that the Hunter's Point site evidenced the formal emergence of an ancient set of practices focusing on ritual curing and power acquisition: the Midewiwin society (1995:33–34).

Engraved copper objects were reported from both the archaeological record and the historic accounts of Lake Superior basin material culture. West noted that copper implements were occasionally decorated with an incised design, which he termed a "lightning emblem" (1929:74, 78). In 1942, Winn published all that he could find about engravings on what he called "decorated coppers." Most of his examples were collected from the surface in Wisconsin. Winn found that copper spear points, crescents, and other implements were often engraved. Sets of parallel dots, zigzag lines, transverse lines, and bird tracks were common motifs (fig. 26; Winn 1942). One implement showed an incomplete ladder motif. Longitudinal rows of indentations in fixed patterns were the most common motif. Most typically they occurred as indentations in one, two, or three parallel lines. In its conventional form of three parallel lines, the two outer lines were symmetrical, with equal numbers of indentations, though the numbers varied from object to object. The central line was generally shorter, offset from the outer lines, and consisted of as few as a single indentation. The indentations themselves appeared to be made with edged tools or punches of metal. Sometimes it was obvious that different tools were used to make variable sets of indentations upon the same object. Decorations occurred most commonly on copper spear points of the rolled socket and rivet hole type; ridge types were rarely decorated. Sometimes the decorations appeared on both sides of an object. Spears and knives showed differing patterns of decoration, and the objects frequently showed use wear. Such depictions on copper and other artifacts were widespread and were found, for example, at Morrison's Island (Kennedy 1966) as well as in Michigan, Minnesota, and to the east in New York and Delaware (Winn 1942:84–85). Winn's explanations of hypothetical meaning were generous in their imaginative range; owner's marks, maker's marks, hunting tallies, poison cavities, decoration for its own sake, and seam bindings were all suggested and evaluated. His conclusions were surprisingly familiar: "The indentures were made on the various implements in accordance with a rule clearly understood and implicitly obeyed" (80). He then made connections with what he termed "ancient magic societies" that may have emerged in the historic period, linking the patterns of indentations with "grades" of achievement sought by members. Another possibility for meaning lies with Phillips's accounts of vision quest behaviors, in which the visionary "might carve images of his guardian spirit on his weapons and on ritual equipment" (1984:25). In this sense the artifacts and their codes may have represented power visions. Warren also documented the occurrence of engraved copper objects from the historic period on the south shore of

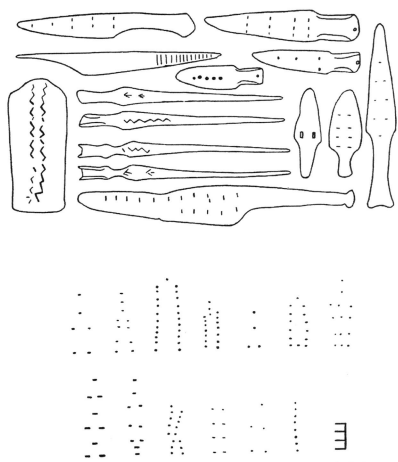

Fig. 26. Decorative designs on prehistoric copper implements from the Lake Superior region. (Winn 1942; reprinted by permission from *Wisconsin Archeologist*.)

Lake Superior. In his report, a local family recorded its genealogy and its claim to significance on "a circular plate of virgin copper, on which is rudely marked indentations and hieroglyphics denoting the number of generations of the family who have passed away since they first pitched their lodges" at LaPointe (1984:89).

Sometimes the icon took actual shape as a copper artifact. Effigy snakes and other symbols of copper were reported from a number of protohistoric sites in the Upper Great Lakes (Brose 1970; Fox 1991; Quimby 1966). Spiral concentric forms (a shape connected in historic times with underwater manitous) were common. The copper spiral was an ornament design discovered in widely separate contexts: prehistoric Wisconsin and Minnesota (Salzer 1974; Stoltman 1973), as well as contact-era eastern U.S. mortuary sites associated both with Iroquoian and Susquehannock peoples (Bradley and Childs 1991; Childs 1994). In the latter case spirals were commonly interred with children. Bradley and Childs (1991) suggested that "spirals and hoops represented an appeal to the healing power of the Panther, a means of invoking protection for those least able to protect themselves" (16).

Historic period iconography also seemed to represent long-lived themes focused on the powers of the manitos. Fox (1992) and Reid (1995) suggested that the ritual breaking of brass dragon side plates on trade guns by Lake Superior basin and northern native peoples represented the ritual killing of the manitou Mishi Bizi. Fox implied that the entire hunting kit was decorated in order to transfer iconographic power to the outcome of the hunt (1992:27). Reid related that copper awls and trade copper cut into triangles were imbued with ritual power for killing or hurting. These examples suggested a widespread and long-lived continuity in the iconography of the native peoples of the Upper Great Lakes that occurred in virtually all media. Likewise, ritual knowledge about the power of copper was both widespread and sought after throughout the Upper Great Lakes in historic, and without a doubt earlier, times. The exact content of that knowledge and the uses to which it was put probably varied from place to place and time to time.

Conclusions

The spatial and contextual distribution of copper and other exotic objects, the social relations that encouraged trade and information exchange, the iconography related to larger mythic meanings, and the cosmologies that ordained that certain things be traded and acquired were all causally related in Upper Great Lakes prehistory. The prehistoric record suggests that the area from the Canadian Shield to the Atlantic coast was linked by ritual behaviors and cosmological beliefs of compatible, or at least articulated, kinds (Hall 1983; Hamell 1987). There may have been limited face-to-face contacts or limited long-distance travel, but through the web of social life and belief, there existed a continuum of behavior and association. In this web of contacts, one person's enemy could be another friend's ally, in the daily or the supernatural world. Alliances could take multiple forms; so could spirits. These things were not always obvious. Space, practicality, and necessity made this complex system possible. Its roots were probably thousands of years deep, like the diplomatic skills of Europeans were. And the stakes were no less high; short- and long-term security, risk reduction, mutual defense, shared territory, personal achievement, and a harmonious cosmos were at issue.

Satisfying the manitous operated as a motivating force for trade among the native peoples of historic and likely earlier times. Trade, in Upper Great Lakes native terms, was tangentially about things, but mostly about mutuality and political intrigue, whether practiced with neighbors, Europeans, or manitous. Manitous were slightly haughty, probably insatiable, and predictably self-important; in short they probably behaved just like human trading partners. Their desires offered a reason to ratchet up and sustain trade. In some prehistoric communities, trade commodities were in demand because they were continually removed from circulation, through burial with the dead and through irretrievable presentation to cosmic powers (Winters 1968). In the case of the Upper Great Lakes, special things, such as tobacco, red cloth, and silver, were left in the country at various shrines and known gathering places of manitous. If the manitous were underwater, the best goods were delivered irretrievably into their territory, the lakes and the rivers. Sometimes this transaction took place with local replaceable objects such as dogs, but more often it took place with goods that apparently had "special value" in the sense used by Winters. Specifically these were goods that were acquired through extended trade between humans. Kohl described such an incident, in which his native informant pitched a bundle of red cloth, tobacco, and silver ornaments over a cliff into the Ontonagon River at night to supplicate the resident manitou for the man's breach of confidence about the locations of copper (1956:63).

These manitous seemed to require that a rightly observant supplicant accumulate, then bury, inundate, immolate, or somehow dedicate a lot of valuable commodities on their behalf (Vecsey 1983). G. A. Wright called this process "ceremonial consumption" (1967:191). Some very important things came to humans in return for these gifts: success in food getting, harmony, well-being, and power. The manitous were both source and destination of important goods. Trade goods purchased food and reduced risk, with the important qualification that the manitous acted as trading partners in the deal. If humans had things that the manitous liked and wanted, which they apparently did, then no more reason was needed to go and get them. In the nineteenth century, the manitous liked things from elsewhere. And they were satisfied thanks to the endless trade that had sustained the ideological and practical well-being of people for millennia. This trade offered ancillary benefits: a chance to see the world, to meet new people, to find a spouse, to learn new songs and dances, to gain some prestige, to acquire powerful things, and to achieve rewards well worth the risks inherent.

In his discussion of prehistoric trade, Earle referred to a process he termed "materialization," meaning the process of identifying a particular material as socially valued. He proposed that "materialization of the abstract social world makes it possible to manipulate the social relationships through control over the economic process. Thus statuses are created and dispensed by those who control the manufacture and circulation of the wealth" (1994:433). What is interesting here is that this statement, for our case, could apply to the manitous as well as to human traders. Not only did one deal within an implicit hierarchy of people who may have managed trade, but with a very obvious hierarchy and specialization of power in the worlds of the spirits. One accompanied the other. The human who had momentary access to the hierarchy of cosmological power gained it to use, in part, as he/she wished, but in constrained settings.[5] Earle concluded that "in egalitarian societies, it is possible to imagine a set of sacred knowledge that is locally available to all initiated individuals . . . by removing the source of that knowledge, access to it becomes more difficult and inherently more easily controlled" (433). But it is not clear that differential access to knowledge leading to segmented control of trade was the case in the prehistoric Upper Great Lakes. In fact at the late end of the scale, the seventeenth century, it appeared that the Lake Superior basin was occupied, or at least visited, by a range of different peoples and that copper acquisition may have occurred with no particular human regulation at all. However, this control probably existed elsewhere, given the contrasts between northern lifeways and Woodland/Mississippian cultural practices further to the south. For instance, perhaps the emergence of multiple sources of copper in the southeast from the twelfth to the fourteenth centuries was caused by differential access to trade and monopolistic controls on traditional copper sources from the north (Goad 1978). The emergence of alternative sources would then allow a way for the ambitious to usurp, challenge, equal, or otherwise reproduce the hegemony of the powerful.

Examining trade arrangements of recent times in the north, it is clear that the region's native people clearly knew how to acquire copper and manipulate its power. But in addition they had a reputation for providing powerful *oki,* or charms, to their trading partners. Some of these partners, for example the Huron, were in awe of northern prowess in the realm of the supernatural. The Huron avidly sought *oki,* for which they paid dearly (Trigger 1987:75). The proof that these *oki* were effective was that they gave the northerners incredible success at the hunt. The Huron connected *oki* with the power to contact spirits controlling uncertainties such as food acquisition, travel safety, and love success. We do not know whether copper was a representation of the Huron concept of *oki.* Copper appeared with some regularity on Huron sites of the late prehistoric and protohistoric period; Fox listed eleven such sites in southwestern Ontario (1991:4). It also occurred on sites in the historic period of their eastern allies, the Susquehannock (Bradley and Childs 1991). It is probable that the actual significance of substances such as copper varied from place to place and people to people. Perhaps the Huron simply passed on copper to eager recipients in exchange for something they themselves prized more fully.

During this complex process of satisfying ones' own and others' demands, everybody ended up with all sorts of their neighbors' material trappings and no doubt shared ideas about their value as well. These exchanges bought good will and ensured future mutual careful treatment and advantage, as well as access to exotic things. For instance, the people of the Upper Great Lakes no doubt could get tobacco from the Huron in exchange for the power of *oki.* Tobacco was something that the northern manitous apparently wanted, something of importance that could not be obtained without the good offices of trading partners elsewhere. This substance had ritual significance to many people. It made an appropriate special value gift to a manitou, one that by its nature involved ceremonial consumption. It also changed form (burned and smoked), which may have made it doubly appropriate.[6]

There is a coherence of environment, behavior, and expressions of meaning within Lake Superior basin cultural groups over long reaches of time. The continuity of the local case, borne out in myth, language, artifacts, iconography of stone-rock-textile-copper, and mortuary behavior was related to the obtaining and maintaining of supernatural power. Copper was precious because it was a medium of power in several ways. Ritually and through the practice of trade, one could cajole the manitous to look after one's affairs. Given this perspective, copper acted simultaneously as a utilitarian and ceremonial substance: in the hunt, in ornamentation, and in negotiations for well-being.[7] The trade links through which copper passed were of ancient origin and linked the Lake Superior basin with other places. The trading system was widely compatible over its spatial extent, despite major cultural and organizational differences at its termini. The needs of the north-south ends of the trading network were compatible, despite the fact that they were in no way identical nor necessarily dependent upon one another.[8]

Suggestions for Further Reading

Baugh, Timothy G. and Jonathon E. Ericson, eds. *Prehistoric Exchange Systems in North America*. New York: Plenum Press, 1994. A comprehensive multi-authored summary of the archaeological and historical evidence for exchange and trade on a continental level.

Bourgeois, Arthur P., ed. *Ojibwa Narratives of Charles and Charlotte Kawbawgam and Jacques LePique, 1893–1895*. Recorded with notes by Homer H. Kidder. Detroit: Wayne State University Press, 1994. Stories and recollections of nineteenth-century Ojibwa life in the changing world of northern Michigan's Lake Superior region.

Johnston, Basil H. *The Manitous: the Spiritual World of the Ojibway*. New York, HarperCollins, 1995. Accounts of Ojibway stories and beliefs from the Georgian Bay area of Ontario.

Rajnovich, Grace. *Reading Rock Art: Interpreting the Indian Rock Paintings of the Canadian Shield*. Toronto, Natural Heritage/Natural History, Inc., 1994. A review of field-based evidence documenting the iconography of the Lake Superior basin and beyond.

8

The Remarkable Enterprise and Acumen of the Natives Are Made Apparent
—WILLIAM H. HOLMES, 1901

Challenges for the Present and Future

> The diversity of invented and exaggerated statements which find currency is, indeed, appalling. The world hears constantly of the discovery of giants and of pygmies; of caverns filled with mummified bodies and rich plunder; of ruined cities abounding in marvelous works of art; of hardened copper; of walls and buildings of astonishing magnitude; of sunken continents; of ancient races associated with extinct species of animals; of inscribed tablets of doubtful origin and extraordinary import.
> —WILLIAM H. HOLMES, 1919

Introduction

Following the saga of prehistoric copper required a long journey through many millennia of North American history. The compilation of the story relied upon the material, archaeological, documentary, and oral historical record to inform, substantiate, and verify general conclusions about the events of the past. This final chapter considers speculative challenges to those observations and conclusions, challenges to the preservation of the region's archaeological deposits, and then discusses the challenges that future archaeological researchers on copper will likely take up. In their efforts to refine and complete the story, copper researchers need the assistance of an informed and active public. The chapter includes suggestions for those whose vocations or avocations might contribute to this collective effort.

The most important conclusions to be gleaned from this volume are these: native peoples achieved a systematic understanding of the geology of copper during their long tenure in the unique environment of the Lake Superior basin; through systematic observation they discovered the places where copper was deposited; their perceptions of their physical environment, like those of people everywhere, were congruent and logically linked with their ideological and cosmological beliefs; and these beliefs were replicated in stories about the causes of change within their world. Nearly two centuries of systematic archaeological research consistently discovered that the exclusive users of Lake Superior copper were none other than the native people of the region, who through direct acquisition as well as trade circulated the metal to other regions of eastern North America.

Mining and metalworking were technologies that required great skill and were developed to a sophisticated level by the indigenous people of the copper-bearing regions. The origins of these technologies lay in stone working and were many millennia old. The technologies were based on systematic observation of natural phenomena, those being the physical properties of stones and metals and their responses to changing conditions such as repeated applications of force and heat.

The use of copper was widespread in prehistory and was an integral part of the adaptations of native peoples to the Lake Superior region from earliest times until the relatively recent appearance of European-influenced cultures. Copper was a component of the material culture of virtually all of the people of the region and was found, in some quantity or another, in the archaeological remains of their households, camps, and other settlements. The importance of copper as a trade material and as a bearer of symbolic meaning was completely compatible with other aspects of Lake Superior basin native material culture and ideologies.

"... Vague Theories, Preconceived Views, and the Demand for Sensational Matter ..."

Given the widely agreed-upon conclusions about the place and importance of copper in indigenous life, it may surprise one to find that another genre of writing discounts these very dearly won facts as accidental and unimportant. This chapter began with a quote from W. H. Holmes ca. 1919, and it provides validation for the old saw that the more things change, the more they stay the same. Despite the facts, demonstrable by archaeology and related sciences, that the early miners were savvy, adaptable, and exclusively indigenous American Indians, erroneous myths about their identities still occasionally surface. The primary myth claims that people from ancient Europe somehow initiated and ran a prehistoric copper industry out of the Lake Superior region. This myth, or one of its closely linked variants, represents some of the most tenacious misinformation in the history of American archaeology. It rests on fantasies that plagued archaeological thinking beginning in the nineteenth century, speculative stories that put untold tons of copper in the hands of technically advanced foreigners from other continents. It derived from the same nineteenth-century racist judgments that propelled the alleged moundbuilder "race" to dizzying heights of "civilization," while reducing living observable Native American people to barbarian status.

Where, precisely, did the mythology of Old World control of North America's past originate? Perhaps its mythic status means that it is by definition not possible to find its origins, nor to explain its tenacity. But something about this myth must be intrinsically satisfying to its proponents. The phenomenon of the copper myth is more than simply an isolated set of unconventional beliefs about local archaeology; it is one example of a wider trend. Others who studied the phenomenon of so-called cult archaeology (Cole 1980) suggested that intertwined and vacillating anti-elitist and authoritarian belief systems were expressed in these myths. Their recurrent popularity was also an indication of an apparent widespread misunderstanding of scientific procedures. Like other cultural constructs, these myths had an internal logic of their own based on intrinsically mutually satisfying but erroneous presumptions about the nature of the contemporary world.

The most prolific and best selling of the myth makers was Barry Fell, whose books *America BC* (1976), *Saga America* (1980), and *Bronze Age America* (1982) earned many dollars and thrilled as well as perplexed their readers. After all, these books sounded as though they were scientific, they appeared to be endorsed by scientists, and Fell himself had general scientific credentials of a sort; he was emeritus professor of marine biology at the esteemed Harvard University. His detailed biography can be found in Williams's wonderful treatise on marginal kooky theorizing, *Fantastic Archaeology* (1991). Fell's stories recounted the discovery of North and South America by Bronze Age Old Worlders. According to Fell, North America was encountered, quite by accident, by Libyans and Iberians on transatlantic drift voyages about 3000 B.C. By 1800 B.C., these accidental tourists were producing copper artifacts, and by 1500–1000 B.C., deliberate voyages from the Old World with the intent of colonization ensued. At some point the Iberians were purportedly joined by the Celts and colonies multiplied.

The material evidence for these explorations, according to Fell, consisted of European-derived copper daggers in North America, stone-carved inscriptions, an evolving Celtiberian culture, ancient markings in a script called Ogam, and observatories, pottery, and other cultural trappings that Fell claimed thrived in New England by 800 B.C. The Celtiberians, he asserted, used the Roman calendar, carved Egyptian and Punic inscriptions, and developed the eastern Algonquian languages (as well as the Zuni culture and the Pima language) and the tumuli cultures of the Adena and Hopewell moundbuilders. In short, according to Fell, the record showed "continuous assimilation of European and Amerindian cultures in the east" (1976:vi–vii).

Fell's books portrayed a worldwide trading network conducted in natural resources and raw materials that thrived between 500–179 B.C. (106–7). Rest assured that Lake Superior native copper played a major role in this trading drama. According to Fell's imaginative schemes, copper, furs, silver, and other raw materials left North and Central America by boat through the Mississippi River, crossed the Atlantic, and were then transformed into trade goods such as tools, weapons, and ornaments in the bronze foundries of North Africa. These trade items were then shipped back to the Americas for eventual distribution and consumption, where they were allegedly found in Hopewell burial mounds and at other localities (96, 127, 128, 165). The Carthaginians helped all this along, horizontally integrating the tin industries of Cornwall, the minerals of North and Central America, the foundries of North Africa, and the lead and leather industries of the Mediterranean in their spheres of exploitation and trade. The North American furs, by the way, went on to India, where their recipient wearers no doubt stayed very warm!

In order to effect this trade, Fell's Carthaginians dealt routinely with the Celts of New England and their confreres, the Celtiberians of central North America. According to Fell, the Celtiberians were already well-established inhabitants of North America, natives even, having pretty much created the materials, cultures, histories, and languages of native North America that archaeology, in its blind ignorance, has to date attributed to American Indians. For despite conventional researches into archaeology, which erroneously attributed all prehistoric archaeological materials to the cultures of American Indians, Fell asserted that there were Libyan ships, Egyptian hieroglyphs, Iberian-Punic inscriptions, and Old World mining ventures well established in the Midwest (4). The voyages ceased after the fall of Rome (479 A.D.), the magic of the North American colonial interlude was forgotten by virtually all, and somehow the presence of these continents was completely expunged from the written records of the Old World.

It is no surprise to learn that conventional archaeology refuted this scheme with real evidence. For example, Fell wrongly identified copper tools as made of the alloy bronze. Yet despite careful, published analyses (Wayman, King, and Craddock 1992), proving without question that North American copper artifacts are not bronze alloys, the Fell myth lives on. The alleged inscriptions and their translations are without evidence according to scholars such as Williams (1991:285). There is equally strong refutation for every single bit of Fell's alleged evidence about Bronze Age use of American copper.

Fell's assertions routinely overlook the critical importance of archaeological context in making interpretations. Suppose, for the moment, that a Bronze Age artifact was reported from a town in Massachusetts. Further suppose that, beyond the name of the town, no one recalled from where the artifact came nor the particular circumstances surrounding its recovery. Upon examination by experts, the artifact looked similar to ones made by people from the British Isles 3,000 or more years old. Now, here's the quandary. Does this mean that the Celts were in North America in 3000 BP? Before you triumphantly say "yes" and decide that buying my book was a waste of money and go buy a few of Barry Fell's instead, you must know more. You must demonstrate that the context (the deposit) from which the object came is congruent with your conclusion. This requires an examination of the means by which the object joined the deposit. Without this information the discovery is meaningless; without it there is no way of determining, for example, whether the Celts dropped a metal ax in 3000 BP or whether a collector abandoned it there last week.

It is true that artifacts from the British Isles occasionally turn up in North America (Fitzhugh and Olin 1993). The whys and hows of these occurrences can only be revealed by the contexts of the deposits that contained them. In the northeast near Boston, these artifacts sometimes occurred in deposits of a particular type. They were probably part of ballast dumps, or places where colonial shippers off-loaded excess ballast, the dead weight they carried in their holds for stability. The ballast, which often consisted of rocks, was dug in known places in England and dumped in many localities near Boston Harbor for three centuries or more (James Bradley, personal communication, 1995). These ballast dumps are classic examples of secondary context, and they can contain all sorts of materials that proclaim and identify their age and origin: rocks of particular English geological origins, coins and ceramics, and a few scattered English antiquities including, occasionally, a Paleolithic flint hand ax.

The important point is that the simple presence of a particular artifact does not indicate that ancient Europeans came to North America. Instead, context demonstrates the high probability that such artifacts got here on a colonial ship with other rocks, a free ride to America as ballast. *It is reasoning from context that builds a factual archaeology.* Artifacts are of great interest, but without context cannot be completely understood. Time and time again Fell's claims to the contrary fail in the face of simpler explanations that take into account archaeological context and formation processes. His myths sell books and apparently thrive through replication with slight modification, as good myths should! There is absolutely no archaeological evidence to date that anyone but indigenous Americans and, subsequent to the seventeenth century A.D., French, British, and Euro-American miners took copper from the Keweenaw or any other mining districts.

Today the long-lived mythology about ancient copper mining recurs in segments of the popular book market. Though this may be no great inconvenience except to the gullible buyer, such myths are in many ways unfortunate. For one thing they are full of inaccuracies. For example, some of these publications announce that prehistory's copper-mining sites were all destroyed by nineteenth-century industrial mining. Believing this to be true, the public may disregard critically important site protection issues. Such stories also encourage site looting rather than emphasize site protection. Finally, some authors confuse speculation with scientific fact, while overlooking the strenuous requirements of the scientific method. The fantasies, being fantasies, are ultimately disappointing, and when this happens, people tend to *blame* archaeology and archaeologists for their loss of innocence. But because the myths do not seem to disappear, it becomes obvious that we must face their apparent immortality and find a way to abide them. In a sense, the myths are of use to the profession of scientific archaeology because they offer a check against the all-too-human tendency to have unwarranted faith in our own infallibility. These myths also provide a way for a wide range of people to become involved and interested in archaeology, admittedly from a weird direction. Over and over again it is clear that people want to participate, to be thrilled by new discoveries, to imagine and create the past, and to get to know its people, and these are all positive trends. There will always be a need to reincorporate vision and participation into our profession without destroying dreams in the process. There is also a pressing need to avoid destroying the sites themselves.

Protecting the Past

Given its global significance, it is particularly important that the archaeological record of copper working be carefully preserved in the Lake Superior basin. The biggest danger, sadly for the integrity of the archaeological deposits of the region, has been the technological refinement of metal detectors, which accelerate the destruction rate of prehistoric sites containing copper. These activities, which involve sensing and then digging up metal artifacts, destroy context and delicate organic constituents of artifacts. This sort of activity is illegal in many places for reasons that are widely valued in our society: the protection of public and private property and the promotion of public over individual determinants of condoned behavior. The effects of metal detecting are felt wherever remains of metal-using cultures are to be found; the scale of this problem in Britain was studied by Dobinson and

Denison (1995). In Michigan, evidence of the predatory character of impacts is apparent in the growth of local collections. In addition, collectors come from all parts of the Midwest to participate in this activity, and their numbers appear to be on the increase.

The scale of the problem in terms of site destruction is impossible to estimate. Anecdotal evidence suggests that the woods are full of people participating in this activity from the earliest days of spring until deep snow and hunting season deter them in the fall. Many are secretive and hide their activities (particularly their successful finds) from one another as well as from archaeologists. Unabashed advertisements, both for tourist-oriented treasure-hunting activities as well as for artifacts on sale, are common in regional and local magazines and newspapers. It is an activity too often marked by competitiveness and disdain for regulation. Property owners and public land stewards find this activity an increasing problem, not only in terms of trespass and liability but also in terms of theft and site destruction. Unfortunately, a coordinated means of promoting change in attitudes about artifact collecting has been slow to occur.

Worst of all, very sensitive data are exposed and destroyed forever. Recall that the question of prehistoric smelting versus colonial smelting could never be resolved, because the relevant artifacts were removed from their original contexts without systematic observation, painstaking record keeping, and pattern interpretation. The artifacts themselves may have survived, but at this point they are somewhat irrelevant. Without their contexts they are mute. Excavation is like a crime scene investigation, or an autopsy, because the act of investigation destroys the objects of interest, which are, for once and for all, not simply artifactual but contextual. Artifacts are a part of archaeological context and to remove them is to devalue them. The precious irreplaceable part of a site is the matrix surrounding an artifact: the soil, the organic remains nearby, the pattern created by soil layering, and sequences of deposition. Learning to read these sediments, deposits, or strata in cross-section is an art as well as a science. Interpreting them is a challenge best met by the combined partnerships of archaeologists with geologists, soils scientists, sedimentologists, and other specialists. Their successful interpretation is the key to understanding events of the past. It is ironic that people ask routinely why archaeology cannot provide all the answers to questions about prehistory. All too often the honest response is simple. Someone who did not care about the complicated procedures of archaeology or did not bother to learn them came along and rubbed out the answers before others could read the fragile record.

Fortunately, there are alternatives to this dismal state of affairs. People who enjoy searching for the past with metal detectors can be assets to archaeological discovery in structured circumstances. Collaborative successes in Britain suggest that similar positive impacts might result in the United States (Dobinson and Denison 1995). Both the closed ranks of the archaeological profession and our tendency to let the public be self-taught about archaeology rather than guided contribute to site destruction problems. In several publicized cases, the collaborative efforts of professional archaeologists and avocational metal detector operators yielded better research results than either group accomplished independently. These sorts of activities should be encouraged and expanded and perhaps will replace destructive artifact hunting. The best known American example of successful collaborative research between metal detectors and archaeologists came from investigations at the Little Bighorn Battlefield National Monument, or, as it is sometimes called, the Custer battlefield (Fox 1993). As Fox succinctly remarked, "From them I learned that a metal detector in any hands does not a metal-detecting machine make. Our project could not have succeeded without these volunteers, and I am more than grateful for their help" (xvi). The potential area of investigation took in nearly 225 hectares. Experienced detector operators swept this vast area with their equipment and systematically flagged metal finds for eventual plotting and excavation. Their work recovered an estimated 20 percent of the actual buried artifacts targeted. By plotting the distribution and individual positions of bullets, adversarial positions were distinguished and the modeling of actual combat phases of the Battle of Little Bighorn resulted. This approach substantially altered the conventional view of the battle's final stages. The story of the final stages of the battle, subject to chaotic contradiction in both written and oral accounts, was ultimately recoverable through systematic archaeological observation, record keeping, and interpretation. Any useful contribution of metal-detecting technology must include these three elements.

There are three themes that avid detector operators and other collectors recurrently mention as important to them regarding archaeology: participation, discovery, and recognition. People with such intense interests should be welcomed into the research stream to build their participation to as great a degree as possible. Much is to be gained by demonstrating that our long-term goals—to discover more about the past—are really the same. Right now our short-term goals and methods are at odds. Promoting education, participation, and collaboration will lead to recognition for collectors who have, in some cases, made archaeology their life's work and avocation. The gain of this meeting of minds and goals will ultimately be to preserve some of what is left of everyone's archaeological record more fully. In fact, it is wonderful to report how excited, involved, and creative the avocationalists can be.

As long as interested people are closed out of the research cycle, both site looting and the appeal of mythical stories will never end. Both the myth makers and the collectors share responsibility for the physical preservation of archaeological sites, but currently that responsibility is being acted out in a tug-of-war mode, an us-against-them frame in which ultimately the physical integrity of the sites is lost. As long as sites and artifacts are seen as places or objects of contention and competition, the data will continue to be ravaged. Therefore, the effective goal for those interested in site preservation must be to find a dimension upon which all the stakeholders agree. Some of those dimensions are local pride and participation in the discovery process. Engaging the myth makers in debate, encouraging collaborative field survey projects, and injecting our professional enterprise with the lay person's abundant enthusiasm and energy will yield positive outcomes. Archaeological professionals can become facilitators and educators in a two-way flow of communication, rather than bestowers of truth.

Efforts to improve communication among some of archaeology's stakeholders are underway across the United States. Many of the most invigorating ideas about the future of collaboration in the Lake Superior basin have come from the many excited avocationalists who live there. For example, a pilot effort to link avocational collectors with teachers and professionals could help educate the youthful public about archaeology. These experts can tour schools and present materials, hands-on, about being responsible for the past. Archaeology displays at state fairs, in which avocational archaeologists communicate their interests to the public, will spread an awareness of prehistory. Such efforts would be a start to circulating valid archaeological information to collectors. Likewise, efforts by archaeology clubs such as state archaeological societies to contact collectors and photograph collections could protect delicate information for the future and could perhaps encourage people to bequeath collections to public institutions.[1] Finally, national organizations of interested stakeholders, composed of professionals, native people, collectors, students, and representatives of state and federal bureaucracies will provide avenues of communication among people with overlapping interests and goals.[2]

The Future of Research

The better and more complete the information about archaeological contexts in the Lake Superior basin and adjacent regions, the more detailed the accounts of past life will become. This work is taking place in the laboratory as well as in the field, but it ultimately depends on rigorous data recovery in both settings. Current research underway in eastern North America is developing finer understandings of alternative copper sources and intercommunity trading patterns (Levine 1996), by attempting to link artifact elemental composition with geographic sources of copper. New data are expanding and correcting the time frames of the earliest copper use and examining more fully its technological and ideological significance to user cultures (Pleger 1996). This work is made possible in part by refinements in radiocarbon technology that now allow dating from tiny amounts of carbon. Work in Michigan now underway looks for the first time at the drainage basin of the Ontonagon River and the acquisition and dispersal of copper through time (Troy Ferone, personal communication, 1996). Other researchers continue the search for metallographic evidence that ancient coppersmiths worked some materials at very high temperatures (S. Terry Childs, personal communication, 1996). Experimental metalworking may play an increasingly important role in establishing the range of prehistoric techniques that were relied upon. Such work can enlighten us about changes in the technological treatment of copper over time and how the earliest technologies probably emerged. Extant collections in museums and in private hands have much to teach us about the range of production techniques as well as the variable forms and functions of copper implements and ornaments. Fragile organics preserved within such collections may yield much information about the ages of artifacts and the methods of their hafting and use. Rigorous documentation of private collections, as well as their loan or donation to public museums and other institutions, will benefit all interested students of prehistory, especially those of the future. The final answers about prehistoric copper working are far from being known; there is exciting work to be done, and no time like now to begin. It will require hard work, painstaking method, wide collaboration, and your participation.

Suggestions for Further Reading

Bahn, Paul. *Archaeology: A Very Short Introduction.* New York: Oxford University Press, 1996. The story of how archaeological problems are posed and solved, in one hundred readable pages.

Dobinson, Colin, and Simon Denison. *Metal Detecting and Archaeology in England.* London: English Heritage and the Council for British Archaeology, 1995. Problems raised and solutions suggested regarding Britain's metal detecting situation.

Feder, Kenneth L. *Frauds, Myths and Mysteries: Science and Pseudoscience in Archaeology.* 2d ed. Mountain View, Calif.: Mayfield Publishing Company, 1996. A brief account of some infamous archaeological hoodwinking, and an explanation of how science and pseudoscience are different.

Henry, Susan L. *Protecting Archeological Sites on Private Lands.* U.S. Department of the Interior, National Park Service, Preservation Planning Branch, Interagency Resources Division. Washington, D.C.: GPO, 1993. Federal legislation protects many archaeological sites on public lands, but ones on private lands are vulnerable without public stewardship. This book offers some protection strategies for sites on private lands.

Purdy, Barbara. *How To Do Archaeology the Right Way.* Gainesville: University Press of Florida, 1996. A guidebook for understanding archaeological reasoning and practicing scientific archaeology, based on examples from Florida.

Vitelli, Karen D., ed. *Archaeological Ethics: Readings from Archaeology Magazine.* Walnut Creek, Calif.: Altamira Press, 1996. An overview of dilemmas archaeologists face: site destruction and looting, repatriation of human remains, relations with living communities, and professional conduct.

Appendix

Some Native Copper Artifacts of the Upper Great Lakes Region

This appendix presents copper tool and ornament forms often associated with archaeological cultures of the Upper Great Lakes and northeastern North America.[1] It emphasizes the cultures of the Archaic and Woodland Traditions. The appendix is arranged alphabetically by tool names that imply general function, but most tools probably functioned in a variety of ways. The range of copper tool forms, shapes, functions, and sizes is vast, and some of the functions of the tools are unknown. Some of this variability is apparent in figure 27.[2] Casual names were given to some of these artifact forms upon their initial discovery. The names derived from European and American hand tools whose functions were assumed to be similar. Early typologies of the region's copper artifacts were prepared by Brown (1904), West (1929), Flaskerd (1940), and Wittry (1957).

Casual names are used as a convenience, but people do not always agree on what an item should be called. One person's "celt" is someone else's "chisel" or "adz." This naming confusion is also true for archaeologists, so it is important to use a unified nomenclature. In the following discussion, a name followed by a Roman numeral indicates the terminology used by Wittry in his descriptive classification of copper tools and ornaments (1957). He divided copper items into ten groups based on generous similarities in assumed function. Within these groups, Wittry assigned specimens to types and subtypes based on similarities in shapes, sizes, and hafting mechanisms. For instance, in Wittry's Group III, Crescents, membership in a type was based on haft element morphology, and membership in a subtype was based on the presence/absence of rivet holes. Wittry's class and type names were general compared to casual names, which often implied specific functions.

Fig. 27. Portions of the John T. Reeder Collection of prehistoric copper artifacts from the Upper Peninsula of Michigan and elsewhere. (Photo courtesy of the MTU Archives and Copper Country Historical Collections, Michigan Technological University.)

APPENDIX

The Wittry classification is the most commonly used framework for describing copper artifacts, but much remains to be learned about the mechanics of artifact production and use, wear patterns, and diverse functions. In the list that follows, cross-references are provided between functional names and those used in the Wittry classification.

All classifications, regardless of their goals, tend to obscure detail. Wittry's classification demonstrates this reality. Some of Wittry's groups were very broad in definition while others were quite specific. Class VIII, spatulas, is an example of a very narrow unit, while Class IV, sometimes called perforators, goes to the opposite extreme. Both Steinbring (1975) and Penman (1977) attempted to refine the Wittry classification; Steinbring incorporated spatial distribution data and manufacturing technology into his revision. Penman studied a set of the same items examined by Wittry, focusing upon the characteristic wear patterns of working edges, and published new information about probable tool functions.

Adz, adze—See Celt

Adzes are wood working tools with a straight or curved working edge, which is sometimes beveled on its inner side. When hafted they lie at right angles to their handles. The working edge of an adze operates perpendicular to the handle and parallel to the line (and presumed grain) of the wood object being worked.

Awl/punch/needle/pike/drill (Wittry Group IV)

This is a very general group which includes tools assumed to have a perforating or piercing function. Awls are long, narrow, general purpose tools of diverse sizes and shapes (figs. 28, 29). They can be round, oval or square in cross section. Their ends can be pointed, hooked, blunted, or flattened. According to Leader (1988:49–52), most awls have at least one pointed end. Some were annealed during manufacture; others were work hardened. Their pointed ends were ground and sometimes resharpened. Some awls were hammered of solid copper bars; others were rolled and compacted from sheet. This category is highly variable in length, from several centimeters to >80 cm (Penman 1977:17). The wide range in size, shape of cross section, and end shape makes it clear that variable function was the rule.

Systematic study of size variation and wear damage to awls suggests that prying, drilling, chipping, and punching were common activities with this implement (Brose 1970:132–33; Pleger 1992:168). Pleger's study revealed that awls and punches were hard to differentiate, but punches were not hafted. Battering and mushrooming on the head of the implement indicated that some sort of force was sometimes transferred to objects by awls. Punched rivet holes in copper tools, for instance, matched the sizes and shapes of some

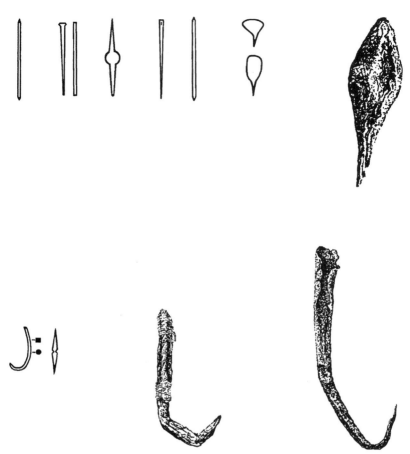

Fig. 28. Top: (l) awl variability (after Wittry 1957; used with permission from *Wisconsin Archeologist*); (r) awl sketch is ca. 9 cm. Bottom: fishhooks: large hook is >8 cm in length; small hook is 3 cm actual size. (Drawing by Brett A. Huntzinger, collections of Michigan Technological University Archaeology Laboratory.)

APPENDIX

Fig. 29. Some awl forms from a private collection, Houghton County, Michigan. (Photo by Patrick E. Martin.)

copper awls, suggesting that awls were used as punches. Many authors concluded that awls were sometimes used as flakers, or tools to produce fine chipping and shaping of stone projectiles (Papworth 1967:205; Ritchie 1965:184).

Based on finds at the Osceola site, Ritzenthaler ("Osceola," 1957:191–92) identified shorter awls as perforators but thought that longer ones (>25 cm) must have functioned in some other way. Both Ritzenthaler and Steinbring (1975:147) commented on the symmetrical appearances of groups of awls found in situ. This uniformity suggested manufacture by the same person, a feature also commented upon by Leader (1988). This symmetry was true of a cluster of 43 awls discovered at 20KE20 (S. Martin 1993). Ritzenthaler suggested that awls appeared in sets of 3–6 items, with some variation in lengths.

Needles are small narrow slender piercers, and sometimes appear with eyes (Kennedy 1966:104). In the Morrison's Island-6 collection their lengths range from 5 to ca. 13 cm. Others are eyeless, and are identical to pins. Pins are short (ca. 2–3 cm), narrow and straight pieces of hammered copper. They probably served a variety of functions: pins, rivets, fish gorges, and needles.

Pikes are very large perforators, ca. > than 20 cm in length. Some pikes measured by Penman, part of the Hamilton collection of copper artifacts at the State Historical Society of Wisconsin, measured as long as 80 cm. Others, reported from the collections of the Minnesota Historical Society and the Field Museum, Chicago, measured 81 cm and 101 cm respectively. Their functions are not perfectly understood. Richard Sackett, speaking about the Minnesota Historical Society collections, offered this conjecture. "The curator of the Society's museum, Willoughby Babcock, believes the Minnesota pike was used in quarrying operations for prying out of rock or mineral masses. Were it used for hunting it would have ruined the pelt of any animal taken with it" (Sackett 1940:55). West thought that these implements, in the 70–100 cm length range, were "used as burning irons in making wooden dugouts, mortars, and other vessels. After burning a series of holes, the splitting of the wood would become an easy task" (1929:93).

Figure 28, top right, shows several forms of this category, including a probable drill variety. Its specific function is not clear to contemporary archaeologists. Such an implement, according to West "reduced to a point at one end and broadened to an inch in width at the other, thus providing a handle that would permit the instrument to be revolved by the fingers when in operation" (1929:96). Kennedy reported that his experimentally produced copper projectile point also worked very well as a drill. "It was found that this point was quite adequate for drilling holes in the fine grained sandstone, even without the aid of a handle or a bow drill arrangement. The point had to

APPENDIX

be resharpened several times during the drilling of each hole; this was done simply by rubbing it on the face of the piece of sandstone being drilled" (1966:107).

Hafting of awls and piercing tools was accomplished in a variety of ways. West described a small awl embedded within a polished bone handle. Penman concluded that awls were intentionally shaped to achieve firm hafting by shaping their widest dimensions in their centers. Some evidence suggested that hafts may have covered as much as half of the length of an awl. Brose suggested that in some cases the pointed end of an awl might have been driven into a handle of bone or wood (1970:133). Sometimes hafted awls appeared as composite tools (Ritchie 1965:184). At Riverside, an awl was recovered in a wooden haft, with a beaver incisor chisel at the opposite end of the handle. The tool "has a 90 degree, pointed bend at the inserted end, a technological attribute elsewhere known in Archaic copper technology" (Hruska 1967:217). The haft was basically a handle with a cut out slot, with a deeper cut at the end of the slot proximal to the butt of the handle, which accommodated a ninety degree bend in the end of the awl. This hook shaped end held the awl in place and prevented it from being dislodged during use. The awl was pressed or driven into the slot, then the hollowed area was filled with a corresponding block of wood, glued and wrapped with rawhide. Some awls known ethnographically had projections on both sides, which provided strength and stability to the haft.

Ax, Axe—See Celt

An ax was a flat chopping or splitting tool hafted at ninety degrees to a handle, whose working edge was parallel to the handle.

Bead (Wittry Group X)

Beads were among the most commonly occurring prehistoric copper objects (fig. 30). Their basic manufacturing sequence was relatively stable throughout prehistory, and they had a very broad geographical distribution. For these reasons it is virtually impossible to determine how old they are without considerable supporting evidence. They were made in one of two ways. Most commonly a thin strip or sheet of copper was rolled or wrapped around itself to form a cylinder or tube with an open center. Other beads appear to have been drilled, but these were not as common as the wrapped form. Beads were produced from small prepared blanks or strips, which were then squeezed or tapped around a circular form to shape the bead. Others were probably made round in a freehand method. Some may have been annealed during the shaping process, but others were left in work hardened condition. These often show cracked or irregular inner curvatures, or yield points. The outer seams of the beads were in some cases carefully burnished or smoothed.

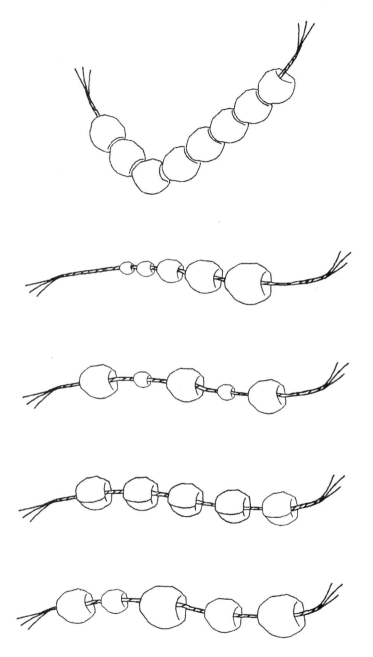

Fig. 30. Some variability in prehistoric bead strands; not to scale. (Drawing by Julia A. Bailey; reprinted, by permission, from S. Martin 1994.)

Very small beads of sheet show, under magnification, that they were made of foil-like layers of copper, probably folded back upon themselves as shaping of the sheet occurred. The bead in figure 22 displays a microstructure which suggested that it was at least partially annealed following initial shaping (Stevens 1996).

The most obvious dimension of variation in beads after mode of manufacture and shape is size, and beads are highly variable. A group of beads of the wrapped cylindrical type varied in length over a range of 1.5 to 9 mm, and 1.9 to 9 mm in diameter (S. Martin 1994). At the Boucher site, beads of the tubular type ranged in length from 10 to 180 mm (Heckenberger, Petersen, and Basa 1990).

Bracelet (Wittry Group IX)

Bracelets, though less common than beads, were reported from a number of prehistoric copper sites and collections (Griffin and Quimby 1961; Kennedy 1966). They were generally made of hammered sheet copper, or of rods of copper (Leader 1988). Bracelets were sometimes decorated with bosses or lines of dots.

Celt/chisel/wedge/gouge/axe/adze (Wittry Group VI)

This group includes implements that vary widely in shape, function, and size. The basic form of these tools is flat, thin, and tapering, with the working edge perpendicular to the long dimension of the tool (figs. 31, 32). Steinbring suggested that their distribution coincided with forested regions and that many of them had functional and formal counterparts in stone (1975:163–65). They are assumed to have been used for wood working.

According to Flaskerd's (1940) defining characteristics, gouges are thin concave wood working tools which typically have a cavity running nearly the length of the inner working side of the tool. West thought that this cavity "may have been intended for the reception of a straight shaft as a handle" (1929:91). He distinguished them from adzes. "The distinguishing feature between an adze and a gouge might be the manner in which the handle is attached, that of a gouge being a straight shaft the same as that of a spud, to be used with both hands, while an adze was probably supplied with a short handle, nearly at right angles to the blade, and used with one hand" (1929:92). Some of these tools probably functioned as splitting wedges or gads (Penman 1977:20). Others were likely handheld and struck on the poll end during use; they were work hardened according to Leader (1988). Many appear to have been heavily battered on the poll end, and were probably used as wood splitters or perhaps as mining implements.

Leader defined chisels as tools that exhibited their widest point at the curved cutting edge. The edge showed chipping from work impacts and

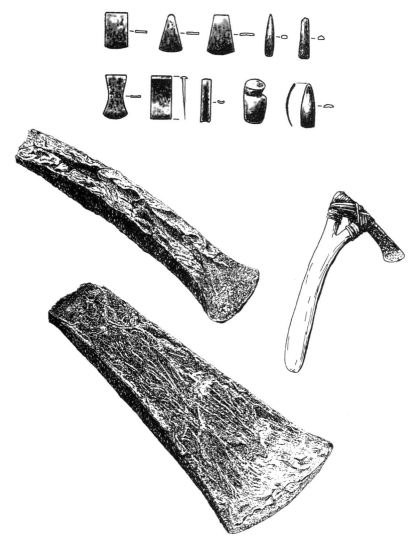

Fig. 31. Top: celt variability (after Wittry 1957; used with permission from *Wisconsin Archeologist*). Center: (l) celt is 17 cm long; (r) hypothetical haft (after Damas 1984:413). Bottom: celt is 26 cm in length. (Drawing by Brett A. Huntzinger, collections of Michigan Technological University Archaeology Laboratory.)

APPENDIX

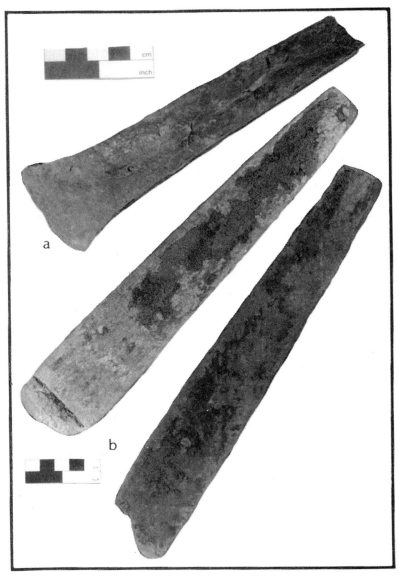

Fig. 32. Some celt forms: (a) found at Bootjack, Michigan; collection of Michigan Technological University Archaeology Laboratory; (b) private collection, Keweenaw Peninsula, Michigan. (Photo by Patrick E. Martin.)

evidence of resharpening. Many were probably handheld and accomplished shaving or graving functions. Typically, they were left in work hardened condition; that is, not annealed after final hammering. According to West they had a broad cutting edge and uniform width, sometimes with a median ridge, and were likely not hafted (1929:88). An example recovered at Summer Island, round and pointed at one end, and the other squared and flat ended, showed wear patterns of short transverse striations on the cutting face, indicating use as a chisel (Brose 1970:133). A similar example was recovered at Chautauqua (Pleger 1992:168). Flaskerd (1940) suggested that chisels functioned as wood working tools for making such things as paddles and canoes, but that they could have been used in many other ways, for instance as choppers to make holes in ice.

Figure 31 (center left) is a celt of unknown age from the Keweenaw Peninsula of Michigan. West commented on the utility of such implements for hafting. "These implements might be classed as chisels, were it not for their being slightly curved from end to end. Hence, when supplied with handles, they could be used without any portion of the handle striking the surface upon which work was being done. These objects either have a square or slightly convex cutting edge and gradually narrow toward the other extremity, which prevents the handle from working toward the cutting edge when in use" (1929:89). Axe hafts were made "by means of a stick of the desired length which had at one end a slot sufficiently deep to receive the ax" (1929:84). Grooved axes provided a seat for increased stability of binding.

Chisel—See Celt

Crescent (Wittry Group III)

Crescents (figs. 33, 34) are arc-shaped items of variable size, form, and function. They have been functionally interpreted as knives, hair ornaments, and gorgets, and are sometimes referred to as semi-lunar knives. These artifacts are so highly variable in size and morphology that there is a lack of full agreement as to their functions. Initially, it appeared that they were comparable to similar forms that appeared in slate and metal among peoples of the sub-Arctic and Arctic culture areas (Flaskerd 1940). Some people identify them by the name *ulu,* borrowed from Inuit "women's knife," which they superficially resemble. Their size variation is noticeable; Steinbring suggested that these implements could vary in size by an order of magnitude (1975:133). Those measured by Penman from the Hamilton Collection varied from 3 to 25 cm in length.

There are many possibilities for handle designs for these implements (fig. 33, bottom). According to Flaskerd (1940:46–47), the haft on the

APPENDIX

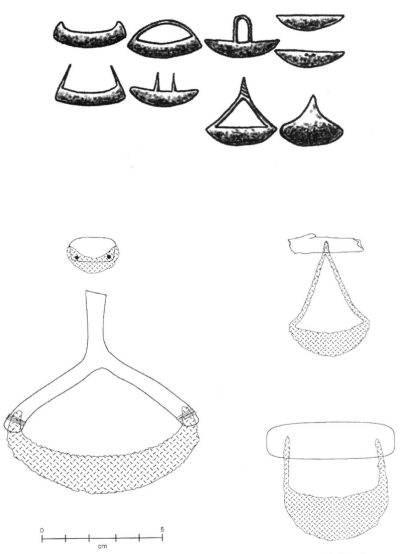

Fig. 33. Top: crescent variability (after Wittry 1957; used with permission from *Wisconsin Archeologist*). Bottom: some hafting possibilities for crescents (Drawing by Julia A. Bailey.)

Fig. 34. Some crescent forms: (a) private collection, Houghton County, Michigan; (b) private collection, Ontonagon County, Michigan. (Photos by [a] Patrick E. Martin; [b] Susan R. Martin.)

prongless varieties probably consisted of a notched piece of bone or wood fitted over the inside blunt curve of the crescent. West suggested that prongless or the so-called canoe shaped crescents were hafted by adding a bone or wood handle. "To make the attachment a groove was doubtless cut in the bone or wood and used for the insertion of the blade" (1929:102). Pronged types likely had wood or bone handles. The center prong forms could be lapped, twisted, or folded as a self handle, or inserted into a wooden or bone grip. "The twisting of slender prongs produces a sharpened, screw-like point which probably was turned into a handle. The wood model for this at Riverside cemetery extended well beyond the point of twisting. Rather than a perpendicular piece, then, the handle might better consist of a shaft of soft wood into which the twisted tang could be inserted" (Steinbring 1975:141). At Riverside, a handle recovered archaeologically was described by Hruska. "Pieces of wood from the handle were visible. It was basically the shape of a slingshot crotch or a 'Y' with one side of the fork fastened to each end of the crescent" (Hruska 1967:197). The wood was identified as birch or maple.

These implements would have been useful for food chopping, bark stripping, or removing hair from hides (Flaskerd 1940). At the Reigh site specimens of the tanged variety had thin cutting edges (Ritzenthaler, "Reigh" 1957:307–8). Pleger stated that intentional beveling and sharpening on the exterior of the arc signified a cutting function (1992:172). Penman examined the edge angles of assumed cutting edges of a collection of crescents (n=22) from Wisconsin surface contexts and compared them to edge angle studies on lithic tools. He concluded that "the great range in edge angles represented in the Hamilton sample would suggest that these were multipurpose tools for butchering meat" (1977:19).

Evidence for use as ornaments exists as well. Incised crescents were quite common, and tangless punctated crescents were recovered at the Reigh site, where they were interpreted as ornaments. Popham and Emerson suggested that small crescents may have functioned as hair ornaments. Others may have been worn as gorgets. Steinbring claimed that they may have had a ceremonial role (1975:141).

Though crescent objects were part of the Middle to Late Archaic Stage core artifact group that defined the Old Copper complex, it has been known since the 1940s that crescentic objects sometimes occurred in Woodland burials (Wilford 1941). They were discovered in association with Laurel and Blackduck pottery in Minnesota (Miles 1951:242) as well as in Hopewell and Adena contexts (John R. Halsey, personal communication, 1996). One occurred in a third century AD context at 20KE20, Michigan. The apparent long time span of dates on crescent objects points to an important reality in archaeology. It is generally context, rather than artifact, that is commonly dated. The best evidence to date points to 1) multiple functions and forms, and 2) multiple ages and long-lived curation for the crescent artifact class. These artifacts exhibit variable wear patterns, variable hafting patterns, variable depositional contexts, and variable temporal contexts.

Drill—See Awl/punch/needle/pike/drill

Fishhook/gorge (Wittry Group VII)

Fishhooks were curved copper rods made in variable sizes and shapes, probably depending on what kinds of fish were to be caught. Some are plain eyeless strips of bent copper rod; others have flattened knobs, notches, or loops for line attachment (Flaskerd 1940). Some have single or multiple barbs, and others have double hooks (Penman 1977:18). Figure 28 (bottom) illustrates a small hook with an elaborately woven line attachment and a larger hook, both from the Keweenaw Peninsula, Michigan.

A gorge is long tapered bar, sometimes with a central indentation, which was suspended from a line and baited to attract fish. Fish gorges were an alternate form of hook, according to Wittry's grouping.

Gad—See Celt

Gaff, Gaffhook

A gaff or gaffhook is a hooked bar used to snag fish from the water. Popham and Emerson reported two such hooks from Ontario, whose total lengths were in the 40–45 cm range (1954:11). Griffin recounted the discovery of several such implements along Lake Superior's north shore (Griffin 1961:104; 109–10). These implements were likely important for capturing large species of fishes.

Gorge—See Fishhook

Gorget—See Miscellaneous ornaments

A gorget is an ornament, sometimes with line holes, which was suspended and worn around the neck or body.

Gouge—See Celt

Harpoon—See Projectile point

Knife (Wittry Group II)

Knives are asymmetrical flat blades with slight to moderate curvature on one edge; this curvature may occur on the working edge, on the backed edge, or both (fig. 35). The working edge is generally beveled on one side. Some knives have two working edges. Knives may be tanged, socketed, or

APPENDIX

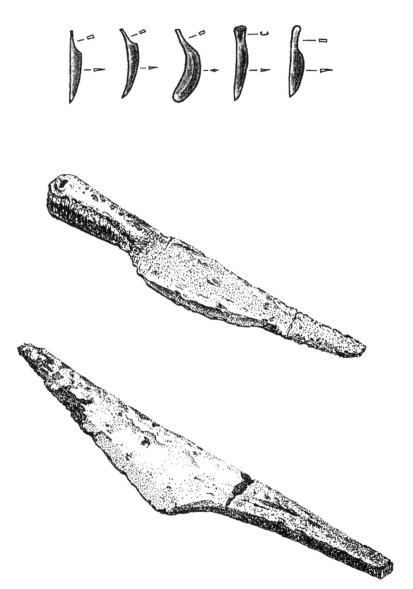

Fig. 35. Top: knife variability (after Wittry 1957; used with permission from *Wisconsin Archeologist*). Center: knife length is <19 cm. Bottom: knife length is >20 cm. (Drawing by Brett A. Huntzinger, collections of Michigan Technological University Archaeology Laboratory.)

appear with a handle like knob; rivet holes are also common in the socketed forms. Some may have a hook on the end of the tang, which probably helped make a secure haft (McPherron 1967:171). They vary in length from 4 to 31 cm.

There is a great deal of variation in the working edges themselves. Some are work hardened and others were annealed. Leader's review of knives in the Wyman collection at Chicago's Field Museum (Leader 1988) documented extensive grinding, work hardening, and hammer packing on the working edges. Wear damage included edge chipping and twisted tips, which Leader concluded was the result of prying and splitting wear. Brose reported that a knife discovered at Summer Island was deliberately blunted on the working edge, implying a special unknown function (1970:133). The association of a copper knife and fish scales recovered from the Lake Superior shoreline implied that such tools could be used for special functions (Anonymous 1978:147–48). Other suggested functions include food and clothing processing as well as wood working. Both Wittry and West reported the presence of decorations on some knives.

Secure hafting was probably very important for the proper functioning of these tools. Both West and Steinbring suggested that the knob end knives probably had leather bindings as grips. West described the probable haft for a straight backed knife as follows: "It has a broad tang, which indicates that the handle was formed by a split stick into which the tang was fitted and then wrapped with rawhide or some other material, to make it secure" (1929:79). Ritzenthaler identified preserved birch wood in association with a socketed knife from the Reigh site ("Reigh" 1957:285).

Miscellaneous ornaments

Copper was hammered into myriad forms which have been interpreted as ornaments; spirals, effigies of animals, pendants, rings, tinkling cones, and perforated discs were typical forms (fig. 36). In some time periods, animal effigy forms were made of copper. Brose (1970) and Quimby (1966) reported copper effigy snakes from protohistoric contexts at the Summer Island and Dumaw Creek sites, respectively. Flaskerd's catalog (1940:48) depicted the range of many forms known from Minnesota. West included rings, decorated strips, conical bangles, circular gorgets, and pendants of varied shapes in his list of prehistoric copper ornaments (1929:111–14).

Needle—See Awl/punch/needle/pike/drill

Perforator, Pick, Pike, Pin—See Awl/punch/needle/pike/drill

Projectile point (Wittry Group I)

Projectile points are the most commonly recovered copper implement in the northern Great Lakes (figs. 37, 38). They were made in a range of

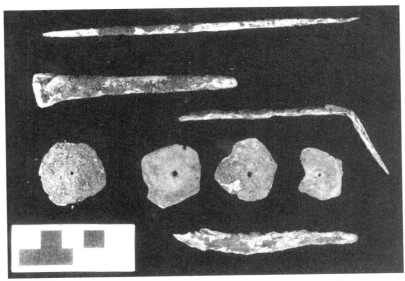

Fig. 36. A cluster of copper ornaments and associated implements, private collection, Keweenaw Peninsula, Michigan. (Photo by Patrick E. Martin.)

forms: flat, conical, beveled, and barbed. Their tangs and sockets took a very diverse range of forms as well, which provided the major distinctions among the many types described. Rat tailed, serrated, notched, expanded, contracting, rounded, and tangless varieties were reported. Socketed forms were sometimes riveted to a handle or shaft, sometimes not. They also varied in size; the smallest conical forms documented by Penman (1977) had a size range of ca. 2 to 9 cm in length. The largest variety, the median ridge socketed forms, ranged from 10 to 29 cm in length. Wittry's original typology included sixteen types of projectiles, but Steinbring (1975) and Penman (1977) concluded that some types could be collapsed and others expanded to depict morphological variation accurately. Both Wittry (1957) and Steinbring (1975) thought that the forms and haft elements of these projectiles probably varied systematically through time and across the region, but provenience problems interfered with conclusions about patterned variation in form. Decorations occurred on the socket side of some socketed forms, and multiple punctations were noted on the blades of some types (Penman 1977:12), Their co-occurrence with use wear suggested that decorations did not distinguish the useful implement from the unused. Edge beveling was very common on some forms.

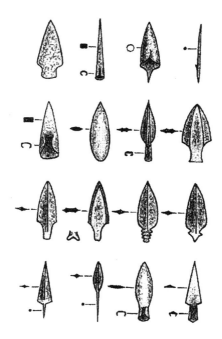
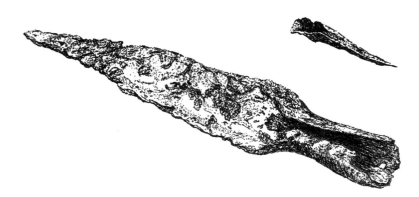

Fig. 37. Top: projectile point variability (after Wittry 1957; used with permission from *Wisconsin Archeologist*). Bottom: small point is ca. 5 cm; large point is ca. 20 cm. (Drawing by Brett A. Huntzinger, collections of Michigan Technological University Archaeology Laboratory.)

APPENDIX

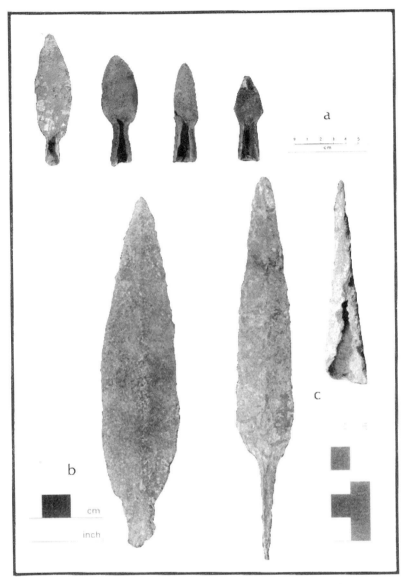

Fig. 38. Some variability in projectile points: (a) private collection, Keweenaw County, Michigan; (b) private collection, Keweenaw Peninsula, Michigan; (c) private collection and the Ontonagon County Historical Society, Ontonagon, Michigan. (Photos by [a] Timothy Pauketat; [b] Patrick E. Martin; [c] Susan R. Martin.)

Leader suggested that the rolled conical forms of projectiles were the simplest; those he inspected at the Field Museum exhibited less than complete annealing (1988:60–61). Socketed forms were produced by a sequence of sheet production, folding and grinding away excess material. His hypothetical sequence for their manufacture follows.

> A copper nugget or laminar plate of sufficient size was hammered, annealed, and ground into a roughly triangular shape. In the case of the spear points with a triangular cross section, the central rib was formed by hammering and then refined by grinding. . . . One side of the triangular copper sheet was chosen as the base. Straight fold lines were carefully hammered into the sheet parallel to each other and perpendicular to the base. The copper corners on either side of the hammered lines were ground off to form parallel edges with the hammered lines. The metal was then very carefully folded along the hammered lines to form the open socket. Shoulders were formed at the junction of the socket top and cutting edge base by hammering the socket folds over and inward at that point. When the hammered metal fold was flush with the surface of the spear point, excess material in the newly created shoulders was ground away. The result is a three sided socket, relatively wide at the base and narrowed at the top. A snug fit was easily obtained for the spear shaft by adjusting the width of the socket through mild hammering or burnishing. (Leader 1988:53–54)

Drill holes or punched holes in the back of the socket served for the attachment of rivets. Sockets for knives were made in an identical fashion.

The wide range of use wear, repeated asymmetrical reworking evidence, and surface polish documented by Penman suggested that projectile points actually served many functions in addition to spearing or stabbing, including prying, gouging, sawing, and wedging/splitting. Leader found evidence, through looking at damage, repairs, and reworking of blades and rivet holes, that thrusting and pulling motions must have been common. He replicated bent and broken tips experimentally by using a spear with an *atl atl* to strike a resistant target. He resolved that annealing actually prolonged the useful life of a projectile point, because hardened ones were brittle and tended to break on impact (1988:56–58). Penman concluded that reworking was common, and that broken projectiles saw reincarnation as knives.

Penman's study of edge angles on projectiles was useful for isolating some of their possible functions. "Hafted knives or "projectile points" with edge angles of 46–55 degrees are multi-purpose tools for cutting sinew, plant fiber shredding, skinning and hide scraping, while those with edges of 66–75 degrees would best serve in wood and bone working and heavy shredding. While many of the Hamilton Collection specimens would fall into these two groups, none fall into a 12–13 degree range which is ideal for whittling wood" (1977:19).

Harpoons represented another special function for projectile points. They occurred in two forms. One was a detachable form that began as a spear tip but was recovered because it attached to a line, presumably embedded in a target fish. The other type was fixed permanently to a shaft, and could be used against a range of fish and aquatic mammals. Pleger (1992:168) reported the discovery of conical two-hole basal-barbed toggling harpoon tips of copper at the Chautauqua Grounds site in eastern Wisconsin. Flat or socketed forms of harpoons and fish spears were multi-barbed, single-barbed, or long or short tang in form (Flaskerd 1940). Multi-barbed harpoons were very toughly made; the backed edges were thick, and some were found with rivet hole(s). West described what he called a notched sort of harpoon: "the notch was doubtless for the attachment of a cord or line to be held in the hand when the shaft or handle was released, in order that the fish, when struck, might be played and controlled by the fisherman" (1929:99).

Hafting mechanisms varied widely for projectile points. Many, particularly the socketed forms, were recovered with organic remains intact within their sockets. Bladdernut wood was documented as shaft material from socketed specimens in the Hamilton Collection (Penman 1977:12). Birch wood shafts were identified at the Reigh site; at Renshaw, species of ash and conifer were noted. At Reigh, archaeological evidence showed that the shaft material extended roughly halfway along the length of the five points recovered there (Ritzenthaler, "Reigh" 1957:307). At Riverside, Spaulding recovered shafts attached to points, "one plainly split to go over tang" (Papworth 1967:163). Steinbring concluded that tanged hafting "is the absolute opposite to socketed hafting, and could reflect historical shifts in available shaft woods. . . . If these tangs are designed to penetrate the shaft end, a pithy center of softwood would be most appropriate. By the same token, hardwoods are better suited to socket hafts" (175:103). He also concluded that some shorter tangs would require mastics and wrapping to make their hafts serviceable.

Punch—See Awl/punch/needle/pike/drill

Spatula (Wittry Group VIII)

Flaskerd and Wittry both used this name to describe a tanged flat wide bladed implement with a rounded or squared end; Wittry suggested that the group was highly variable in morphology (fig. 39 top left, fig. 40a). Though their functions were not understood completely, Flaskerd proposed that they were used as fish scalers, knives, scrapers, or pottery smoothers (1940:46).

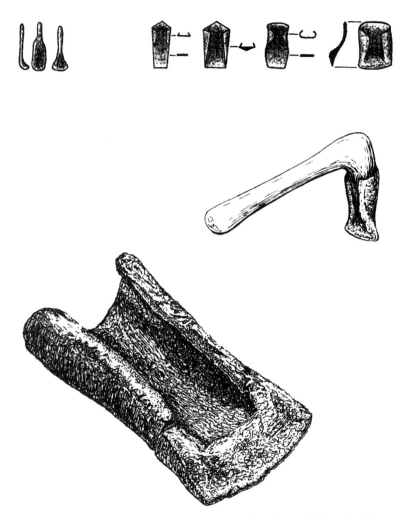

Fig. 39. Top: (l) spatula and (r) spud variability (after Wittry 1957; used with permission from *Wisconsin Archeologist*). Center: hypothetical hafted spud (after Ritzenthaler 1957:280). Bottom: spud length is 13 cm. (Drawing by Brett A. Huntzinger, collections of Michigan Technological University Archaeology Laboratory.)

APPENDIX

Fig. 40. (a) Spatula and (b) spuds from a private collection and Ontonagon County Historical Society, Ontonagon County, Michigan. (Photo by Susan R. Martin.)

Some, he suggested, were hafted, while others were used without handles. They ranged in size from ca. 12–16 cm in length.

Spear—See Projectile point

Spud (Wittry Group V)

Spuds were massive cutting and chopping tools, hafted with the working edge parallel to the handle, with a ninety degree angle at the haft. They varied in shaped and could be flat, bell shaped, grooved/notched, and double bitted; some were socketed (fig. 39 top right, center, bottom; fig. 40b). Penman's study of edge angles suggested that they were likely used for working wood and bone (1977:19). Flaskerd added uses for debarking trees, chopping holes in ice, and producing dugouts (1940:41). Steinbring concluded that spuds were widely distributed and were frequent objects of trade. Occasionally, like many other utilitarian objects, they were decorated. One such specimen was a spud incised with a continuous diamond shaped pattern, recovered in Houghton County, Michigan (Cleland and Wilmsen 1969:26–27).

Figure 39 (center) illustrates a hypothetical haft for a socketed spud. This arrangement was proposed by Ritzenthaler ("Reigh" 1957:280), who was inspired by such finds at both the Reigh and Osceola sites in Wisconsin. West proposed that handles for spuds were driven deeply, and secured by rawhide and pitch. The spud socket "ends in a shoulder against which the end of the shaft would solidly rest. This shoulder or offset brings the face of the blade on an even plane with the edge of the shaft, thus enabling the operator to cut to a line and at right angles to any depth" (West 1929:83).

Wedge—See celt

A wedge was a narrow object probably used for splitting wood or perhaps mining (Flaskerd 1940:42)

Notes

CHAPTER 2

1. The trap range refers to the areal extent of the bedrock in which veins of native copper are found.
2. Henrich's description of the Ducktown, Tennessee, copper ore deposit is very interesting because it reveals the discovery, in about 1880, of archaeological finds alleged to indicate aboriginal smelting of copper ore. As the original source is rather rare, it is included in full text below. The account states that the finder, Judge James Parks of Ducktown, discovered a number of aboriginal artifacts (what he called moundbuilder pottery) on the surface of an old field after a downpour, and "to his surprise, however, he found also, besides these Indian relics, a few pieces of rich carbonate copper-ore and pieces of slag or cinder, besides many fragments of pots, which appeared to have been lined with a charred mass. Some of the slag-pieces were of peculiar shape, as if they had cooled adhering to the upper rim of a crucible, after the main contents of the crucible had been poured out" (1895:176). Other remains reportedly included a "small irregular slab of metallic copper" and "parts of artificially-cut square fire-proof stones of oblong shape, (one of them glazed and sintered on its head from the effect of fire)" (176). Henrich assessed the ores at Ducktown and some of the recovered artifacts, concluding that

> we have here, apparently, a perhaps unique remnant of a former metallurgical industry, that of melting copper from its oxidized ores, practiced by the mound-builders, who probably melted the richest carbonate ores, occurring in an easily-melting, self-fluxing ferruginous gangue, in small crucibles made of a peculiar but easily-cut rock of the neighborhood, and used for this purpose a furnace built of oblong rectangular brick cut from the same kind of rock as the crucible. This is the surmise of the writer from the relics of the industry seen by him. As Judge Parks is in correspondence with the Smithsonian Institution with regard to this matter, it is to be hoped that a thorough search of the spot by excavation will secure, before it is too late, additional and comprehensive relics of a pre-historic industry. (176–77)

Henrich is to be commended for his perceptive recognition of the recurrent tragedy of site destruction. As he correctly inferred, the question of whether the smelter evidence was in fact associated with aboriginal remains can only be settled by meticulous excavation. Collecting the materials and removing them from context contributed to the destruction of the site and lessened the prospect for clear resolution of questions of association.

3. The turbidity and color described by Schoolcraft are the result of the downcutting path of the Ontonagon River through the clay lake bed of postglacial Lake Ontonagon.
4. In the modern day, the belief that the lake controls short- and longer-term climatic conditions is repeated in a commonly heard cliché about Isle Royale: "This island makes its own weather."

Chapter 3

1. Some refer to it as the New York Ethnological Society.
2. Time-space systematics can be defined as the establishment of a place in time and space for artifacts and the development of a taxonomy of archaeological cultures based on these coordinates. Squier and Davis had no comparative frame of reference for arranging their finds in relative order.
3. "Invading" is their word (Squier and Davis 1848:6).
4. The forestry data were attributed to William Henry Harrison, U.S. president and a budding archaeological thinker (Squier and Davis 1848:305–6).
5. The Law of Superposition borrows from geological reasoning. In an undisturbed deposit, lower strata predate those above.
6. Michael Wayman and colleagues examined that portion of Squier and Davis's metals collection held by the British Museum (Wayman, King, and Craddock 1992) and again demonstrated that all copper artifacts attributable to central U.S mounds were native copper rather than introduced metals.
7. "Minesota" is the historically correct spelling of the name of the mine.
8. They were not alone. Fred Dustin put his critique of Ferguson kindly: "Perhaps his enthusiasm misled him somewhat" (Dustin 1957:7).
9. West (1859–1938) would have been sixteen years old in 1875.
10. There is no immediate evidence of a familial link between the two Messrs. West. Roy Owen West (1868–1958), born in Georgetown, Illinois, was a well-known lawyer in the Chicago area and had, in the same week that the McDonald-Massee expedition left Milwaukee, been appointed secretary of the interior by President Coolidge. George Armor West (1859–1938) was the only son of a family from Racine County, Wisconsin, a lawyer, and a trustee of the Milwaukee Public Museum from 1906 until 1938.
11. This institution is now Michigan Technological University.
12. This work is now underway in the Ontonagon River drainage (Troy Ferone, personal communication, 1996).

Chapter 4

1. Agricola also expressed some traditional concerns about supernatural influences in mines. According to his experiences, the underground presence of "demons of ferocious aspect" was a real problem in medieval mines (Hoover and Hoover 1950 [VI]:217–18). Agricola actually wrote a book about such beasts, *De Animatibus Subterraneis,* and prescribed that some could be expelled "by prayer and fasting." The really stubborn ones, however, could not be moved, and the only solution was to shut the mine. He rated stubborn demons as fifth on a list of seven reasons why mines were sometimes abandoned.
2. The nineteenth-century companies apparently considered mining this drift for profit. "In digging cellars, constructing roads, and exploring trenches, such pieces are so common, that it has been thought that they would pay for their collection by washing the earth" (Whittlesey 1863:15).
3. Collection of drift (a.k.a. float) copper occurred sporadically across the entire range of glacial drift deposition in central North America, as far from bedded sources as Iowa, Indiana, Illinois, Ohio, and Nebraska. Rickard calculated that copper was sporadically available as a constituent of glacial drift over an area of 450,000 square miles (1934:273).
4. Fox was a bit prone to underestimate labor demands; he once stated that there was a whole summer's worth of archaeology to be done at Isle Royale (1911:100).
5. In a letter dated September 1, 1897, on letterhead from the Office of the Superintendents of Poor of Ontonagon County, is this brief description of provenance: "Michigan State Geological Survey. 41 ancient hammers and fragments from Caledonia bluff. (3 Hammers and 30 fragments purchased from Dr. W. C. Gates Rockland Mich included in the lot" (Letter on file, Ontonagon County site information file, Archaeology Laboratory, Michigan Technological University). The signature is that of local businessman and prominent community member Benjamin F. Chynoweth (1851–1916), captain of the National Mine and brief owner of the Mass Mine, as well as Ontonagon County supervisor and county surveyor.
6. This was certainly the case in the nineteenth century when the area of the southern mainland trap range was the territory of the Ontonagon band, which occupied a series of villages at the mouth of the local river(s) and occupied the interior at will and in the winter.

Chapter 5

1. The continuities between lithic technologies and those of copper working have been examined more recently by Jack Steinbring (1975, 1991). He wished to demonstrate that the prehistoric copper industries of central North America adopted formal attributes in tool shape from earlier Paleoindian lithic industries (1975:25, 28–29).

2. Decades ago, Willoughby experimented with such an alloy (algodonite), heating it to white-hot temperatures to drive off volatile arsenic and render it easier to work. Ruhl called this the first experimental evidence that North American Indians may have been involved in real metallurgy (Greber and Ruhl 1989:143).
3. Michael LaRonge had an interesting result.

> During my experiment I had great success in maintaining a wood fire temperature of between 865–920 degrees Celsius. At its peak it reached 1073 degrees Celsius. Somewhat hotter than other citations you found. My fire pit was roughly 63 cm across with C-shaped berm around three-quarters of it to take advantage of a light wind cycling through the courtyard in which the experiment took place. The C-shape was mainly to allow access to the base of the fire for heating the raw copper. No air blowers were used, human or otherwise. I began with pine kindling and shortly there after went strictly to oak. After starting the fire I let it build up a good bed of coals over an hour and kept adding wood every fifteen or so minutes from then on, as I recall. The measure device I used was a pyrometer from a sculpture studio used for large-scale casting of metals. Of course it may have been off, but I do not believe so due to the necessity of exact temperature reading during that process. (Michael LaRonge, personal communication, 1997)

4. A collection, after all, is not an instance of primary deposition or in situ context, nor is it an archaeological site or assemblage. Provenance problems with items in collections are all too common and are the bane of the existence of museum curators, whose most frequent laments, after hand-wringing about lack of funds, are "Where did that come from?" and "Why didn't anyone keep better records?"

CHAPTER 6

1. Radiocarbon dates, expressed as age in radiocarbon years, are statistical estimates. They are accompanied by a set of qualifier numbers that need some explanation. Radiocarbon dates are averages and include a known error range (standard deviation) that describes how consistent the individual measures that make up the average are. The more consistent the individual measures, the smaller the standard deviation. Adding the standard deviation to both sides of the average date enlarges the probability that the date is correct (i.e., falls within the stated plus/minus range two times out of three). Doubling the standard deviation further increases confidence that the real date falls within the stated range. Some dates also include a calibration correction figure (cal) that reflects adjustments for known fluctuations in atmospheric carbon levels.
2. Other collections may contain such data as well; many of those copper items collected casually in the Upper Peninsula include organic remains within haft elements. Well-meaning people often clean out the remains of the hafts from the sockets of copper tools and discard them as waste. This is a mistake. Such remains are important and should be protected.
3. Nipissing refers to the ancient shoreline of a postglacial lake that filled the Lake Superior basin around 5500–4500 years ago. The highest elevation of the water

NOTES TO CHAPTER 7

levels in this lake are calculated to be about 198 m asl at the Keweenaw Peninsula (Saarnisto 1975:302). Algoma refers to the final established postglacial water levels that preceded modern levels. Algoma levels were a bit higher than modern ones in the western Lake Superior basin, and date to about 2000 BP.

4. For example, there are legendary stories about OCC sites overrun as paper mills, power plants, and logging operations altered the deltas of the Escanaba, Michigan, region.
5. To get a rough estimate of the calendar age of a date expressed in radiocarbon years, subtract the date from 1950. For example, a date of 1800 BP yields a rough uncalibrated estimate of A.D. 150; a date of 3000 BP becomes 1050 B.C.
6. It is difficult for those of us who learned archaeology after the advent of radiocarbon dating to appreciate just what a revolution in thinking this breakthrough gave to the discipline of archaeology. According to the authors' carefully worded account from 1957, it was not yet clear that fluted points predated the OCC in Wisconsin, and they were not willing to conjecture that they were (Wittry and Ritzenthaler 1957:323).
7. "Initial Woodland" refers to the first pottery-using peoples of the region. The pots themselves are distinct from those of earlier southern pottery users, so "Early Woodland" is reserved for southerly pottery making archaeological cultures. In the Upper Great Lakes, Initial Woodland and Middle Woodland refer to the same time period, ca. 2500–1500 BP.
8. The term "ingot" as used by Salzer does not refer to a melted or smelted product, but to a hammered regular shape of worked copper, a blank to be worked into a formal artifact.
9. The Copper Country Archives at Michigan Technological University include the photographic collection of several copper collections of late nineteenth-century vintage from the western Upper Peninsula. In one of the glass plates documenting the collection of John T. Reeder, a similar spiral is identified from Two Rivers, Wisconsin (fig. 27).

CHAPTER 7

1. "Middlesex" refers to a cultural manifestation found in New York defined by Ritchie and Dragoo (1960) and known primarily from burial evidence. "Meadowood" is an early Woodland phase associated with pottery, also from New York (Ritchie 1944).
2. Compare, for example, Fogel's distributions with those of Goodman (1984).
3. The native residents of the prehistoric and protohistoric Lake Superior basin belonged to a number of linked ethnic groups who shared, to some degree, similar languages, adaptations, and ideological beliefs.
4. The spellings of manitou names are as given in Bourgeois (1994).
5. Crumley referred to the special quality of potential but variable ranking as heterarchy, "the relation of elements to one another when they are unranked or when they possess the potential for being ranked in a number of different ways"

(1995:3). This heterarchy is a useful beginning for understanding the situational or ad hoc qualities of Upper Great Lakes native social and political leadership and its cosmological parallels. Leaders and their manitou counterparts possessed differential power in alternative circumstances and acted to maximize this power situationally. But the upper hand or power monopoly was always temporary. For example, when it came to hunting success, the Lake Superior people could access power from a manitou such as Mishi Ginabig, but they also had to supplicate the particular owner or boss of the animal sought. Vecsey put it this way. "The Ojibwas chose leaders for specific tasks; one person's leadership never transcended all of the community's needs. The same situation applied to the various owners; each Owner served a purpose but had no control over activities beyond its scope of powers" (1983:76).
6. The transformative aspect of tobacco was important. Allouez related that during a tobacco sacrifice, people solemnly chanted and vocalized as the tobacco turned into a column of smoke and maintained their attention to it until it dispersed (Kellogg 1917:112).
7. Trevelyan's research (1987) considers that the manipulation of copper was an essential revitalization ritual throughout eastern United States prehistory.
8. Yet there appeared to be compelling continuities in some shared aspects of prehistoric spiritual life. For instance, the wearing of the famous copper-clad antlers in ritual context at Hopewell would have effected, at least to a Lake Superior native person, a very powerful appearance reminiscent of the Mishi Ginabig. A number of authors have commented on these parallels (Willoughby 1935).

CHAPTER 8

1. One such organization is the Michigan Archaeological Society, PO Box 359, Saginaw, Michigan 48606. Another is the Wisconsin Archeological Society, PO Box 1292, Milwaukee, Wisconsin 53201.
2. There are a number of regional and national efforts to engage avocational archaeologists in educational and research settings. Many states in the United States conduct a yearly "Archaeology Week" or "Archaeology Month," which helps to educate citizens as well as to promote awareness about site preservation issues. Interested persons can contact his/her state's historic preservation office for more information. National organizations such as the Society for American Archaeology (SAA) provide information for teachers to incorporate archaeological data within primary and secondary school curricula. For further information, contact the SAA organizational office at 900 Second Street NE #12, Washington, DC 20002–3557. They may be reached via the internet at http://www.saa.org.

APPENDIX

1. For coverage of copper artifact forms produced and used by cultures of the Lower Great Lakes, the Mississippi Valley, and the American southeast, see

Greber and Ruhl (1989) for Hopewell materials, Leader (1988) for Hopewell and Mississippian/southeast copper artifacts, and Trevelyan (1987) for Hopewell and Mississippian/southeast artifacts from an art historian's perspective.

2. John T. Reeder's vast collection was primarily composed of Upper Great Lakes copper specimens. Mr. Reeder was a resident of Houghton County, Michigan, from 1889 until his death in about 1937. After his death some of his collection was donated to the Cranbrook Institute of Science in Bloomfield Hills, Michigan. The whereabouts of the rest is unknown.

References

Anderton, John B. *Paleoshoreline Geoarchaeology in the Northern Great Lakes, Hiawatha National Forest, Michigan.* Hiawatha National Forest Heritage Program Monograph 1. Escanaba, Mich.: Hiawatha National Forest, 1993.

Anderton, John B., and Walter L. Loope. "Buried Soils in a Perched Dunefield as Indicators of Late Holocene Lake-level Change in the Lake Superior Basin." *Quaternary Research* 44 (1995): 190–99.

[Anonymous]. Our Lake Superior Trip, III. *University Magazine* [Ann Arbor], April 1869.

[Anonymous]. "The Photo Album: A Copper Knife from Lake Superior." *Wisconsin Archeologist* 59, no. 1 (1978): 147–48.

Avouris, Dulcinea. "Notes on Isle Royale and Caledonia-area Hammerstones." Unpublished ms. on file at the Archaeology Laboratory, Michigan Technological University, Houghton, 1996.

Baerreis, David A., Hiroshi Daifuku, and James E. Lundsted. "The Burial Complex of the Reigh Site, Winnebago County, Wisconsin." *Wisconsin Archeologist* 38, no. 4 (1957): 244–78.

Barnouw, Victor. *Wisconsin Chippewa Myths and Stories and Their Relation to Chippewa Life.* Madison: University of Wisconsin Press, 1977.

Bastian, Tyler. "Prehistoric Copper Mining in Isle Royale National Park, Michigan." Master's thesis, Department of Anthropology, University of Utah, Salt Lake City, 1963.

———. "Trace Element and Metallographic Studies of Prehistoric Copper Artifacts in North America: A Review." In *Lake Superior Copper and the Indians: Miscellaneous Studies of Great Lakes Prehistory,* ed. James B. Griffin, pp. 151–75. Anthropological Papers 17. Ann Arbor: Museum of Anthropology, University of Michigan, 1961.

Baugh, Timothy G., and Jonathon E. Ericson, eds. *Prehistoric Exchange Systems in North America.* New York: Plenum, 1994.

Beaubien, Paul L. "Report on Archeological Investigations at Isle Royale NP. Beaubien to Regional Historian, Region 2, Omaha, Nebraska." Copy of letter of July 9, 1953, on file, Isle Royale National Park headquarters, Houghton, Michigan.

Beukens, R. P., et al. "Radiocarbon Dating of Copper-preserved Organics." *Radiocarbon* 34, no. 3 (1992): 890–97.
Binford, Lewis R. "Archaeology as Anthropology." *American Antiquity* 28, no. 2 (1962): 217–25.
Bird, Junius B. "The 'Copper Man': A Prehistoric Miner and His Tools from Northern Chile." In *Pre-Columbian Metallurgy of South America: A Conference at Dumbarton Oaks, October 18th and 19th, 1975,* ed. Elizabeth P. Benson, pp. 105–32. Washington, D.C.: Dumbarton Oaks Research Library and Collections, 1979.
Bisbing, Richard, and Susan R. Martin. "Analysis of Fiber and Textiles from 20KE20." *Michigan Archaeologist* 39, nos. 3, 4 (1993): 170–75.
Bornhorst, Theodore J. "Tectonic Context of Native Copper Deposits of the North American Midcontinent Rift System." In *Middle Proterozoic to Cambrian Rifting, Central North America,* ed. R. W. Ojakangas, A. B. Dickas, and J. C. Green, pp. 127–36. Geological Society of America Special Paper 312. Boulder: Geological Society of America, 1997.
Bornhorst, Theodore J., and William I. Rose. "Self-guided Geological Field Trip to the Keweenaw Peninsula, Michigan." *Proceedings of the Institute on Lake Superior Geology* 40, no. 2, 1994.
Bourgeois, Arthur P., ed. *Ojibwa Narratives of Charles and Charlotte Kawbawgam and Jacques LePique, 1893–1895,* recorded with notes by Homer H. Kidder. Detroit: Wayne State University Press, 1994.
Bourque, Bruce J. "Evidence for Prehistoric Exchange on the Maritime Peninsula." In *Prehistoric Exchange Systems in North America,* ed. Timothy G. Baugh and Jonathon E. Ericson, pp. 23–46. New York: Plenum, 1994.
———. "Maine State Museum Investigation of the Goddard Site, 1979." *Man in the Northeast* 22 (1981): 3–27.
Bradley, James, and S. Terry Childs. "Basque Earrings and Panther's Tails: The Form of Cross-cultural Contact in Sixteenth Century Iroquoia." In *Metals in Society: Theory beyond Analysis,* ed. Robert Ehrenreich. *MASCA Research Papers in Science and Archaeology* 8, no. 2 (1991): 7–17.
Brose, David S. *The Archaeology of Summer Island: Changing Settlement Systems in Northern Lake Michigan.* Anthropological Papers 41. Ann Arbor: Museum of Anthropology, University of Michigan, 1970.
———. "Trade and Exchange in the Midwestern United States." In *Prehistoric Exchange Systems in North America,* ed. T. G. Baugh and J. E. Ericson, pp. 215–40. New York: Plenum, 1994.
Brown, Charles E. "Myths, Legends, and Superstitions about Copper." *Wisconsin Archeologist* 20, no. 2 (1939): 35–40.
———. "The Native Copper Implements of Wisconsin." *Wisconsin Archeologist* 3, no. 2 (1904): 49–98.
Buchner, A. P., and L. F. Pettipas. "The Early Occupations of the Glacial Lake Agassiz Basin in Manitoba; 11,500 to 7,700 B.P." In *Archaeological Geology of North*

References

America, ed. N. P. Lasca and J. Donahue, pp. 51–59. Centennial Special Volume 4. Boulder: Geological Society of America, 1990.

Buckmaster, Marla M., and James R. Paquette. "The Gorto Site: A Preliminary Report on a Late Paleoindian Site in Marquette County, Michigan." *Wisconsin Archeologist* 69, no. 3 (1988): 101–24.

Childs, S. Terry. "Native Copper Technology and Society in Eastern North America." In *Archaeometry of Pre-Columbia Sites and Artifacts: Proceedings of a Symposium Organized by the UCLA Institute of Archaeology and the Getty Conservation Institute*, ed. David A. Scott and Pieter Meyers, pp. 229–53. Los Angeles: Getty Conservation Institute, 1994.

Clark, Caven Peter. *Archeological Survey and Testing at Isle Royale National Park, 1986–1990 Seasons*. Lincoln, Neb.: Midwest Archeological Center, National Park Service, 1990.

———. *Archeological Survey and Testing, Isle Royale National Park, 1987–1990 Seasons*. Occasional Studies in Anthropology 32. Lincoln, Neb.: Midwest Archeological Center, National Park Service, 1995.

———. "Group Composition and the Role of Unique Raw Materials in the Terminal Woodland Substage of the Lake Superior Basin." Ph.D. diss., Department of Anthropology, Michigan State University, 1991.

———. "Plano Tradition Lithics from the Upper Peninsula of Michigan." *Michigan Archaeologist* 35, no. 2 (1989): 89–112.

Clark, David E., and Barbara A. Purdy. "Early Metallurgy in North America." In *Early Pyrotechnology: The Evolution of the First Fire-using Industries*, ed. Theodore A. Wertime and Steven F. Wertime, pp. 45–58. Washington, D.C.: Smithsonian Institution Press, 1982.

Cleland, Charles E. "Comments on 'A Reconsideration of Aboriginal Fishing Strategies in the Northern Great Lakes Region.'" *American Antiquity* 54, no. 3 (1989): 605–9.

———. "The Inland Shore Fishery of the Northern Great Lakes: Its Development and Importance in Prehistory." *American Antiquity* 47, no. 4 (1982): 761–84.

———. *The Prehistoric Animal Ecology and Ethnozoology of the Upper Great Lakes Region*. Anthropological Papers 29. Ann Arbor: Museum of Anthropology, University of Michigan, 1966.

———., ed. *The Lasanen Site: An Historic Burial Locality in Mackinac County, Michigan*. Anthropological Series 1. East Lansing: Michigan State University Museum, 1971.

Cleland, Charles E., Richard D. Clute, and Robert E. Haltiner. "Naub-Cow-Zo-Win Discs from Northern Michigan." *Midcontinental Journal of Archaeology* 9, no. 2 (1984): 235–49.

Cleland, Charles E., and Edwin N. Wilmsen. "Three Unusual Copper Implements from Houghton County, Michigan." *Wisconsin Archeologist* 50, no. 1 (1969): 26–32.

Coghlan, H. H. "Metallurgical Analysis of Archaeological Materials." *Viking Fund Publications in Anthropology* 28 (1960): 1–20.

———. "Native Copper in Relation to Prehistory." *Man* 51, no. 156 (1951): 90–93.

———. "A Note upon Native Copper: Its Occurrence and Properties." *Proceedings of the Prehistoric Society* 3 (1962): 58–67.

Cole, John R. "Cult Archaeology and Unscientific Method and Theory." In *Advances in Archaeological Method and Theory III*, ed. Michael B. Schiffer, pp. 1–33. New York: Academic, 1980.

Conway, Thor, and Julie Conway. *Spirits on Stone: The Agawa Pictographs*. Heritage Discoveries Publication 1. San Luis Obispo, Calif.: Heritage Discoveries, 1990.

Craddock, B. "The Experimental Hafting of Stone Mining Hammers." In *Early Mining in the British Isles, Proceedings of the Early Mining Workshop, Plas Tan y Bwlch, November 1989*, ed. P. and S. Crew. [Plas Tan y Bwlch], Wales: Early Mining Workshop, 1990.

Craddock, Paul T. *Early Metal Mining and Production*. Washington, D.C.: Smithsonian Institution Press, 1995.

Craig, Alan K., and Robert C. West. *In Quest of Mineral Wealth: Aboriginal and Colonial Mining and Metallurgy in Spanish America*. Geoscience and Man 33. Baton Rouge: Louisiana State University, 1994.

Crumley, Carole L. "Heterarchy and the Analysis of Complex Societies." In *Heterarchy and the Analysis of Complex Societies*, ed. R. M. Ehrenreich, C. L. Crumley, and J. E. Levy, pp. 1–5. Archeological Papers 6. Arlington, Va.: American Anthropological Association, 1995.

Cunningham, Wilbur M. *Study of the Glacial Kame Culture in Michigan, Ohio, and Indiana*. Occasional Contribution 12. Ann Arbor: Museum of Anthropology, University of Michigan, 1948.

Cushing, Frank H. "Primitive Copper Working: An Experimental Study." *American Anthropologist* 7 (1894): 93–117.

Damas, David. "Copper Eskimo." In *Handbook of North American Indians, Volume 5: Arctic*, ed. William C. Sturdevant, pp. 397–414. Washington, D.C.: Smithsonian Institution, 1984.

Dawson, K. C. A. "Cummins Site: A Late Palaeo-Indian (Plano) Site at Thunder Bay, Ontario." *Ontario Archaeology* 39 (1983): 3–31.

Dice, Lee R. *The Biotic Provinces of North America*. Ann Arbor: University of Michigan Press, 1943.

Dobinson, Colin, and Simon Denison. *Metal Detecting and Archaeology in England*. London: English Heritage and the Council for British Archaeology, 1995.

Donaldson, William S., and Stanley Wortner. "The Hind Site and the Glacial Kame Burial Complexes in Ontario." *Ontario Archaeology* 59 (1995): 5–95.

Draper, Lyman C. "Fabrication of Ancient Copper Implements." *Collections of the State Historical Society of Wisconsin, Being a Page-for-page Reprint of the Original Issue of 1879* 8 (1908): 165–73.

Drexler, C. W., W. R. Farrand, and J. D. Hughes. "Correlation of Glacial Lakes in the Superior Basin with Westward Discharge Events from Lake Agassiz." In *Glacial Lake Agassiz*, ed. J. T. Teller and L. Clayton, pp. 310–29. Special Paper 26. St. John's, Newfoundland: Geological Association of Canada, 1983.

Drier, Roy W. "Archaeology and Some Metallurgical Investigative Techniques." In *Lake Superior Copper and the Indians: Miscellaneous Studies of Great Lakes Prehistory,* ed. James B. Griffin, pp. 134–47. Anthropological Papers 17. Ann Arbor: Museum of Anthropology, University of Michigan, 1961.

Drier, Roy W., and Octave J. Du Temple, eds. *Prehistoric Copper Mining in the Lake Superior Region.* Calumet, Mich., and Hinsdale, Ill.: N.p., 1961.

Dudzik, Mark J. "First People: The Paleoindian Tradition in Northwestern Wisconsin." *Wisconsin Archeologist* 72, nos. 3, 4 (1991): 137–54.

Dustin, Fred. "An Archaeological Reconnaissance of Isle Royale." *Michigan History* 41 (1957): 1–34.

Earle, Timothy. "Positioning Exchange in the Evolution of Human Society." In *Prehistoric Exchange Systems in North America,* ed. Timothy G. Baugh and Jonathon E. Ericson, pp. 419–37. New York: Plenum, 1994.

Ellis, H. Holmes. "Flint-working Techniques of the American Indians." Unpublished ms., Department of Archaeology, Ohio State Museum, Columbus, 1940.

Ericson, Jonathon E. "Toward the Analysis of Lithic Production Systems." In *Prehistoric Quarries and Lithic Production,* ed. Jonathon E. Ericson and Barbara A. Purdy, pp. 1–9. Cambridge: Cambridge University Press, 1984.

Eschman, D. E., and P. F. Karrow. "Huron Basin Glacial Lakes: A Review." In *Quaternary Evolution of the Great Lakes,* ed. P. F. Karrow and P. E. Calkin, pp. 79–93. Special Paper 30. St. John's, Newfoundland: Geological Association of Canada, 1985.

Farrand, W. R., and C. W. Drexler. "Late Wisconsinan and Holocene History of the Lake Superior Basin." In *Quaternary Evolution of the Great Lakes,* ed. P. F. Karrow and P. E. Calkin, pp. 17–32. Special Paper 30. St. John's, Newfoundland: Geological Association of Canada, 1985.

Fell, Barry. *America B.C.: Ancient Settlers in the New World.* New York: Quadrangle/New York Times Book Company, 1976.

———. *Bronze Age America.* Boston: Little, Brown, 1982.

———. *Saga America.* New York: Time Books, 1980.

Ferguson, William P. F. "The Franklin Isle Royale Expedition." *Michigan History Magazine* 8 (1924): 450–68.

———. "Michigan's Most Ancient Industry: The Prehistoric Mines and Miners of Isle Royale." *Michigan History Magazine* 7 (1923): 155–62.

Fitting, James E. "Middle Woodland Cultural Development in the Straits of Mackinac Region: Beyond the Hopewell Frontier." In *Hopewell Archaeology: The Chillicothe Conference,* ed. David S. Bose and N'omi Greber, pp. 109–12. Kent, Ohio: Kent State University Press, 1979.

Fitzgerald, William, and Camille Ramlukan. "Accessing the Supernatural: Algonkian Devotional Items from the Hunter's Point (BfHg-3) Site." *Arch Notes, Newsletter of the Ontario Archaeological Society* 2 (1995): 8–17.

Fitzgerald, William R., and P. G. Ramsden. "Copper-based Metal Testing as an Aid to Understanding Early European-American Interaction: Scratching the Surface." *Canadian Journal of Archaeology* 12 (1988): 153–61.

Fitzhugh, William W., and Jacqueline S. Olin. *Archeology of the Frobisher Voyages.* Washington, D.C.: Smithsonian Institution Press, 1993.

Flaskerd, George. "A Schedule of Classification, Comparison, and Nomenclature for Copper Artifacts in Minnesota." *Minnesota Archaeologist* 6, no. 2 (1940): 35–50.

Fogel, Ira L. "The Dispersal of Copper Artifacts in the Late Archaic Period of Prehistoric North America." *Wisconsin Archeologist* 44, no. 3 (1963): 129–80.

Ford, Richard, and David S. Brose. "Prehistoric Wild Rice from the Dunn Farm Site." *Wisconsin Archeologist* 56, no. 1 (1975): 8–15.

Foster, John W., and Josiah D. Whitney. *Report on the Geology and Topography of a Portion of the Lake Superior Land District in the State of Michigan. Part 1: Copper Lands.* U.S. House. 31st Cong., 1st sess, 1850. H. Doc. 69.

Fox, George R. "The Ancient Copper Workings on Isle Royale." *Wisconsin Archeologist* 10, no. 2 (1911): 72–[108].

———. "Isle Royale Expedition." *Michigan History* 13, no. 2 (1929): 308–23.

Fox, Richard A., Jr. *Archaeology, History, and Custer's Last Battle: The Little Big Horn Reexamined.* Norman: University of Oklahoma Press, 1993.

Fox, William A. "Dragon Sideplates from York Factory: A New Twist on an Old Tail." *Manitoba Archaeological Journal* 2, no. 2 (1992): 21–35.

———. "The Serpent's Copper Scales." *Wanikan, Newsletter of the Thunder Bay Chapter of the Ontario Archaeological Society* 91-03 (1991): 3–15.

Fox, William A., R. G. V. Hancock, and L. A. Pavlish. "Where East Meets West: The New Copper Culture." *Wisconsin Archeologist* 76, nos. 3, 4 (1995): 269–93.

Frank, Leonard. "A Metallographic Study of Certain Pre-Columbian American Implements." *American Antiquity* 17, no. 1 (1951): 57–59.

Franklin, Ursula. "Folding: A Prehistoric Way of Working Native Copper in the North American Arctic." *MASCA Journal* 2, no. 2 (1982): 48–52.

Garrad, Charles. "Thoughts about the Hunter's Point (BfHg-3) Site." *Arch Notes, Newsletter of the Ontario Archaeological Society* 95, no. 3 (1995): 32–35.

Gillman, Henry. "Ancient Works at Isle Royale, Michigan." *Appleton's Journal*, August 9, 1873, pp. 173–75.

Glumac, Petar D. "Recent Trends in Archaeometallurgical Research: Introduction." *MASCA Research Papers in Science and Archaeology* 8, no. 1 (1991): 5–6.

Goad, Sharon I. "Chemical Analysis of Native Copper Artifacts from the Southeastern United States." *Current Anthropology* 21, no. 2 (1980): 270–71.

———. "Exchange Networks in the Prehistoric Southeastern United States." Ph.D. diss., Department of Anthropology, University of Georgia, 1978.

———. "Middle Woodland Exchange in the Prehistoric Southeastern United States." In *Hopewell Archaeology: The Chillicothe Conference,* ed. David S. Brose and N'omi Greber, pp. 239–46. Kent: Kent State University Press, 1979.

Goad, Sharon I., and John Noakes. "Prehistoric Copper Artifacts in the Eastern United States." In *Archaeological Chemistry II,* ed. G. Carter, pp. 335–46. Advances in Chemistry Series 171. Washington, D.C.: American Chemical Society, 1978.

Goodman, Claire G. *Copper Artifacts in Late Eastern Woodlands Prehistory.* Evanston, Ill.: Center for American Archaeology, Northwestern University, 1984.
Greber, N'omi, and Katherine Ruhl. *The Hopewell Site: A Contemporary Analysis Based on the Work of Charles C. Willoughby.* Boulder: Westview Press, 1989.
Green, William, James B. Stoltman, and Alice B. Kehoe, eds. "Introduction to Wisconsin Archaeology: Background for Cultural Resource Planning." *Wisconsin Archeologist* 67, nos. 3, 4 (1986): 163–395.
Griffin, James B. "Hopewell and the Dark Black Glass." *Michigan Archaeologist* 11 (1965): 115–55.
———., ed. *Lake Superior Copper and the Indians: Miscellaneous Studies of Great Lakes Prehistory.* Anthropological Papers 17. Ann Arbor: Museum of Anthropology, University of Michigan, 1961.
Griffin, James B., and George I. Quimby. "The McCollum Site, Nipigon District, Ontario." In *Lake Superior Copper and the Indians: Miscellaneous Studies of Great Lakes Prehistory,* ed. James B. Griffin, pp. 91–102. Anthropological Papers 17. Ann Arbor: Museum of Anthropology, University of Michigan, 1961.
Hack, John T. "Post-glacial Drainage Evolution and Stream Geometry in the Ontonagon Area, Michigan." *Professional Paper 504-B.* Washington, D.C.: United States Geological Survey, 1965.
Hall, Robert L. "A Pan-continental Perspective on Red Ochre and Glacial Kame Ceremonialism." In *Lulu Linear Punctated: Essays in Honor of George Irving Quimby,* ed. Robert C. Dunnell and Donald K. Grayson, pp. 74–107. Anthropological Papers 72. Ann Arbor: Museum of Anthropology, University of Michigan, 1983.
Hamell, George R. "Mythical Realities and European Contact in the Northeast during the Sixteenth and Seventeenth Centuries." *Man in the Northeast* 33 (1987): 63–87.
Hancock, R. G. V., et al. "Distinguishing European Trade Copper and Northeastern North American Native Copper." *Archaeometry* 33 (1991): 69–86.
Hansel, A. K., et al. "Late Wisconsinan and Holocene History of the Lake Michigan Basin." In *Quaternary Evolution of the Great Lakes,* ed. P. F. Karrow and P. E. Calkin, pp. 39–53. Special Paper 30. St. John's, Newfoundland: Geological Association of Canada, 1985.
Harrison, Christina, et al. *The Paleo-Indian of Southern St. Louis County, Minnesota: The Reservoir Lakes Complex.* Dubuque, Iowa: Kendall/Hunt, 1995.
Heckenberger, Michael, James Petersen, and Louise Basa. "Early Woodland Period Ritual Use of Personal Adornment at the Boucher Site." *Annals of the Carnegie Museum* 59, no. 3 (1990): 173–217.
Heckenberger, Michael, et al. "Early Woodland Period Mortuary Ceremonialism in the Far Northeast: A View from the Boucher Cemetery." *Archaeology of Eastern North America* 18 (1990): 109–44.
Hedican, Edward J., and James McGlade. "A Taxometric Analysis of Old Copper Projectile Points." *Man in the Northeast* 45 (1993): 21–38.
Henrich, Carl. "The Ducktown Ore Deposits and the Treatment of the Ducktown

Copper-ores." *Transactions of the American Institute of Mining Engineers* 25 (1895): 173–245.

Hill, Mark A. *Ottawa North and Alligator Eye: Two Late Archaic Sites on the Ottawa National Forest*. Cultural Resource Management Series Report Number 6. Ironwood, Mich.: Ottawa National Forest, 1994.

———. "The Timid Mink Site: A Middle Woodland Domestic Structure in Michigan's Western Upper Peninsula." *Wisconsin Archeologist* 76, nos. 3, 4 (1995): 338–64.

Holman, Margaret B., and Terrance J. Martin, eds. "The Sand Point Site (20BG14)." *Michigan Archaeologist* 26, nos. 3, 4 (1980): 1–90.

Holmes, William H. "Aboriginal Copper Mines of Isle Royale, Lake Superior." *American Anthropologist* 3 (1901): 684–96.

———. *Handbook of Aboriginal American Antiquities, Part 1: Introductory, the Lithic Industries*. Bureau of American Ethnology Bulletin 60. Washington, D.C.: Government Printing Office, 1919.

Hoover, Herbert C., and Lou Henry Hoover, eds. *Georgius Agricola: De Re Metallica, Translated from the First Latin Edition of 1556*. New York: Dover, 1950.

Hoy, P. R. "Fabrication of Ancient Copper Implements." *Collections of the State Historical Society of Wisconsin, Being a Page-for-page Reprint of the Original Issue of 1879* 8 (1908): 169–73.

Hruska, Robert. "The Riverside Site: A Late Archaic Manifestation in Michigan." *Wisconsin Archeologist* 48, no. 3 (1967): 145–260.

Hurst, Vernon, and Lewis Larson. "On the Source of Copper at the Etowah Site, Georgia." *American Antiquity* 24, no. 2 (1958): 177–81.

Janaway, R. C. "Dust to Dust: The Preservation of Textile Materials in Metal Artefact Corrosion Products with Reference to Inhumation Graves." *Science and Archaeology* 27 (1985): 29–34.

Janzen, Donald. *The Naomikong Point Site and the Dimensions of Laurel in the Lake Superior Basin*. Anthropological Papers 36. Ann Arbor: Museum of Anthropology, University of Michigan, 1968.

Johnston, Basil. *The Manitous: The Spiritual World of the Ojibway*. New York: HarperCollins, 1995.

Jovanovic, Borislav. "The Origins of Metallurgy in South-east and Central Europe and Problems of the Earliest Copper Mining." In *The Origins of Metallurgy in Atlantic Europe, Proceedings of the Fifth Atlantic Colloquium*, ed. Michael Ryan, pp. 335–43. Dublin: Stationery Office, 1978.

Jury, Wilfrid W. "Copper Artifacts from Western Ontario." *Wisconsin Archeologist* 46, no. 4 (1965): 223–46.

———. "Copper Cache at Penetanguishene, Ontario, Canada." *Wisconsin Archeologist* 54, no. 2 (1973): 84–106.

Kellogg, Louise P. *Early Narratives of the Northwest, 1634–1699*. New York: Charles Scribner's Sons, 1917.

Kennedy, Clyde C. "Preliminary Report on the Morrison's Island-6 Site." *National Museum of Canada Bulletin* 206 (1966): 100–125.

King, Mary E. "Textile Fragments from the Riverside Site, Menominee, Michigan." *Verhandlungen des XXXVIII Internationalen Amerikanisten-kongresses* 1 (1968): 117–23.

Kohl, Johann G. *Kitchi-Gami: Wanderings Round Lake Superior.* Minneapolis: Ross and Haines, 1956.

LaBerge, Gene L. *Geology of the Lake Superior Region.* Phoenix: Geoscience Press, 1994.

Landes, Ruth. *Ojibwa Religion and the Midewiwin.* Madison: University of Wisconsin Press, 1968.

Lanman, Charles. *A Canoe Voyage up the Mississippi and around Lake Superior in 1846.* Grand Rapids, Mich.: Black Letter, 1978.

Larsen, Curtis E. "Lake Level, Uplift, and Outlet Incision, the Nipissing and Algoma Great Lakes." In *Quaternary Evolution of the Great Lakes,* ed. P. F. Karrow and P. E. Calkin, pp. 63–77. Special Paper 30. St. John's, Newfoundland: Geological Association of Canada, 1985.

Leader, Jonathan M. "Technological Continuities and Specialization in Prehistoric Metalwork in the Eastern United States." Ph.D. diss., Department of Anthropology, University of Florida, Gainesville, 1988.

Levine, Mary Ann. "Native Copper, Hunter-gatherers, and Northeastern Prehistory." Ph.D. diss., Department of Anthropology, University of Massachusetts, 1996.

———. "Tracing the Exchange of Native Copper among Hunter-gatherers in Northeastern Prehistory." Paper presented at the 60th annual meeting of the Society for American Archaeology, Minneapolis, May 3–7, 1995.

Libby, W. F. *Radiocarbon Dating.* 2d ed. Chicago: University of Chicago Press, 1955.

Lovis, William A., et al. "Environment and Subsistence at the Marquette Viaduct Locale of the Fletcher Site." In *Investigating the Archaeological Record of the Great Lakes State,* ed. Margaret B. Holman, Janet G. Brashler, and Kathryn E. Parker, pp. 251–305. Kalamazoo: New Issues Press, Western Michigan University, 1996.

Maddin, R., T. Stech Wheeler, and J. D. Muhly. "Distinguishing Artifacts Made of Native Copper." *Journal of Archaeological Science* 7 (1980): 211–25.

Martin, Susan R. "An Archaeological Site Examination of the Gros Cap Cemetery Area, Mackinac County, Michigan—Phase II." *Cultural Resource Management Report 6.* Houghton: Department of Social Sciences, Michigan Technological University, 1979.

———. "Michigan Prehistory Facts: The State of Our Knowledge about Ancient Copper Mining in Michigan." *Michigan Archaeologist* 41, nos. 2, 3 (1995): 119–38.

———. *Models of Change in the Woodland Settlement of the Northern Great Lakes Region.* Ann Arbor: University Microfilms, 1985.

———. "A Possible Bead Maker's Kit from North America's Lake Superior Copper District." *Beads: Journal of the Society of Bead Researchers* 6 (1994): 49–60.

———. "A Reconsideration of Aboriginal Fishing Strategies in the Northern Great Lakes Region." *American Antiquity* 54, no. 3 (1989): 594–604.

———., ed. "20KE20: Excavations at a Prehistoric Copper Workshop." *Michigan Archaeologist* 39, nos. 3, 4 (1993): 127–93.

Martin, Terrance J. "Prehistoric Animal Exploitation on Isle Royale." In *Archeological Survey and Testing, Isle Royale National Park, 1987–1990 Seasons*, by Caven P. Clark, pp. 205–16. Occasional Studies in Anthropology 32. Lincoln, Neb.: Midwest Archeological Center, National Park Service, 1995.

Mason, Carol I., and Ronald J. Mason. "The Age of the Old Copper Culture." *Wisconsin Archeologist* 42, no. 4 (1961): 143–55.

Mason, Ronald J. *Great Lakes Archaeology*. New York: Academic, 1981.

———. "The Paleo-Indian Tradition." In *Introduction to Wisconsin Archaeology: Background for Cultural Resource Planning*, ed. James B. Stoltman, William Green, and Alice B. Kehoe. *Wisconsin Archeologist* 67, nos. 3, 4 (1986): 181–206.

Mason, Ronald J., and Carol Irwin. "An Eden-Scottsbluff Burial in Northeastern Wisconsin." *American Antiquity* 26, no. 1 (1960): 43–57.

McHale Milner, Claire. "Regional Identity and Interregional Interaction during the Juntunen Phase, A.D. 1200–1620." Paper presented at the Canadian Archaeological Association conference, Edmonton, Alberta, 1994.

McPherron, Alan. *The Juntunen Site and the Late Woodland Prehistory of the Upper Great Lakes Area*. Anthropological Papers 30. Ann Arbor: Museum of Anthropology, University of Michigan, 1967.

Merrill, George P. *The First One Hundred Years of American Geology*. New York: Hafner, 1924; rpt., 1967.

Miles, Suzanne W. "A Revaluation of the Old Copper Industry." *American Antiquity* 16, no. 3 (1951): 240–47.

Moffat, Charles R., and Janet M. Speth. "Rainbow Dam: Late Archaic and Middle Woodland Habitation in the Wisconsin River Headwaters." Paper presented at the 61st annual meeting of the Society for American Archaeology, New Orleans, 1996.

National Park Service. Letter file, reference A 9015-ISRO, Drier to McLaughlin, March 11, 1953. Isle Royale National Park Headquarters, Houghton, Michigan.

Noble, Bruce J., Jr., and Robert Spude. "Guidelines for Identifying, Evaluating, and Registering Historic Mining Properties." *National Register Bulletin* 42 (1992): 1–30.

Nordeng, Steve. "A Magnetic Gradient Survey of 20KE20, Keweenaw County, Michigan." *Michigan Archaeologist* 39, nos. 3, 4 (1993): 145–47.

O'Brien, William. *Mount Gabriel: Bronze Age Mining in Ireland*. Galway: Galway University Press, 1994.

Packard, R. L. "Pre-Columbian Copper-Mining in North America." Annual Report of the Board of Regents of the Smithsonian Institution. 52nd Cong., 2nd sess. *Miscellaneous Document* 114, no. 1 (1893): 175–98.

Papworth, Mark L. *Cultural Traditions in the Lake Forest Region during the Late*

High-water Stages of the Post-glacial Great Lakes. Ann Arbor: University Microfilms, 1967.

Pauketat, Timothy. "Lithic Analysis." *Michigan Archaeologist* 39, nos. 3, 4 (1993): 176–80.

Penman, John. "The Old Copper Culture: An Analysis of Old Copper Artifacts." *Wisconsin Archeologist* 58, no. 1 (1977): 3–23.

Pettipas, Leo. "Recent Developments in Paleo-Indian Archaeology in Manitoba." *Archaeological Survey of Alberta Occasional Paper* 26 (1985): 39–63.

Phillips, Ruth B. "Dreams and Designs: Iconographic Problems in Great Lakes Twined Bags." *Bulletin of the Detroit Institute of Arts* 62 (1986): 27–37.

———. *Patterns of Power: The Jasper Grant Collection and Great Lakes Indian Art of the Early Nineteenth Century.* Kleinburg, Ont.: McMichael Canadian Collection, 1984.

Pleger, Thomas C. "A Functional and Temporal Analysis of Copper Implements from the Chautauqua Grounds Site (47-Mt-71), a Multicomponent Site Near the Mouth of the Menominee River." *Wisconsin Archeologist* 73, nos. 3, 4 (1992): 160–76.

———. "The Red Ochre Complex: A Series of Related Cultures Exhibiting Shared Ceremonial Mortuary Patterns during the Late Archaic to Early Woodland Transition: 1,200 B.C.–A.D. 1." Unpublished paper in possession of the author, 1996.

Pletka, Karyn. "Chemical Trace Element Analysis of Native Copper." *Michigan Archaeologist* 37, no. 4 (1991): 283–97.

Popham, Robert E., and J. N. Emerson. "Manifestations of the Old Copper Industry in Ontario." *Pennsylvania Archaeologist* 24 (1954): 1–19.

Porter, James. "Hixton Silicified Sandstone: A Unique Lithic Material Used by Prehistoric Cultures." *Wisconsin Archeologist* 42 (1961): 78–85.

Purdy, Barbara A. "Pyrotechnology: Prehistoric Application to Chert Materials in North America." In *Early Pyrotechnology: The Evolution of the First Fire-using Industries,* ed. Theodore A. Wertime and Steven F. Wertime, pp. 31–44. Washington, D.C.: Smithsonian Institution Press, 1982.

Quimby, George I. "The Dumaw Creek Site: A Seventeenth Century Prehistoric Indian Village and Cemetery in Oceana County, Michigan." *Fieldiana: Anthropology* 56, no. 1 (1966): 1–91.

Rajnovich, Grace. *Reading Rock Art: Interpreting the Indian Rock Paintings of the Canadian Shield.* Toronto: Natural Heritage/Natural History, 1994.

Rapp, George, Eiler Hendrickson, and James Allert. "Native Copper Sources of Artifactual Copper in Pre-Columbian North America." In *Archaeological Geology of North America,* ed. N. Lasca and J. Donahue, pp. 479–98. Centennial Special Volume 4. Boulder: Geological Society of America, 1990.

Reid, C. S. "Fur Trade 'Consumer' Site Assemblage Patterning: Expectations and Actuality." In *Proceedings of the First Historical Archaeology Conference of the Upper Midwest (HACUM), Redwing, Minnesota,* ed. John P. McCarthy and Phyllis E. Messenger. Minneapolis: Institute for Minnesota Archaeology, 1995.

Renfrew, Colin. *The Emergence of Civilization: The Cyclades and the Aegean in the Third Millennium B.C.* London: Methuen, 1972.

Richner, Jeffrey S. "Depositional History and Stone Tool Industries at the Winter Site; A Lake Forest Middle Woodland Cultural Manifestation." Master's thesis, Department of Anthropology, Western Michigan University, 1973.

Rickard, T. A. "The Use of Native Copper by the Indigenes of North America." *Journal of the Royal Anthropological Institute of Great Britain and Ireland* 64 (1934): 265–87.

Ritchie, William A. *The Archaeology of New York State.* Garden City, N.Y.: Natural History Press, 1965.

———. "The Pre-Iroquoian Occupations of New York State." Memoir 1. Rochester: Rochester Museum of Arts and Sciences, 1944.

Ritchie, William A., and Don W. Dragoo. "The Eastern Dispersal of Adena." Bulletin 379. Albany: New York State Museum and Science Service, 1960.

Ritzenthaler, Robert E. "The Osceola Site, an Old Copper Site Near Potosi, Wisconsin." *Wisconsin Archeologist* 38, no. 4 (1957): 186–203.

———. "Reigh Site Report—Number 3." *Wisconsin Archeologist* 38, no. 4 (1957): 278–310.

Ritzenthaler, Robert E., ed. "The Old Copper Culture of Wisconsin." *Wisconsin Archeologist* 38, no. 4 (1957): 185–329.

Ritzenthaler, Robert E. et al. "The Pope Site, a Scottsbluff Cremation in Waupaca County." *Wisconsin Archeologist* 53, no. 1 (1972): 15–19.

Ritzenthaler, Robert E., and George I. Quimby. "The Red Ochre Culture of the Upper Great Lakes and Adjacent Areas." *Fieldiana: Anthropology* 36, no. 11 (1962): 243–75.

Ritzenthaler, Robert E., and Warren L. Wittry. "The Oconto Site—An Old Copper Manifestation." *Wisconsin Archeologist* 38, no. 4 (1957): 222–44.

Root, William C. "Metallurgical Examination of Five Copper Artifacts from Southern Michigan." In *Lake Superior Copper and the Indians: Miscellaneous Studies of Great Lakes Prehistory,* ed. James B. Griffin, pp. 148–50. Anthropological Papers 17. Ann Arbor: Museum of Anthropology, University of Michigan, 1961.

Ross, William. "The Interlakes Composite: A Re-definition of the Initial Settlement of the Agassiz-Minong Peninsula." *Wisconsin Archeologist* 76, nos. 3, 4 (1995): 244–68.

Rothenberg, Beno, and Antonio Blanco-Freijeiro. *Studies in Ancient Mining and Metallurgy in South-west Spain: Explorations and Excavations in the Province of Huelva.* London: Institute for Archaeo-Metallurgical Studies, Institute of Archaeology, University of London, 1981.

Ryan, Michael, ed. *The Origins of Metallurgy in Atlantic Europe, Proceedings of the Fifth Atlantic Colloquium.* Dublin: Stationery Office, 1978.

Saarnisto, Matti. "Stratigraphical Studies on the Shoreline Displacement of Lake Superior." *Canadian Journal of Earth Sciences* 12 (1975): 300–319.

Sackett, Richard R. "Unusual Copper Artifacts in Collection of the Minnesota Historical Society." *Minnesota Archaeologist* 6, no. 2 (1940): 54–55.

Salisbury, R. D. "Notes on the Dispersion of Drift Copper." *Wisconsin Academy of Sciences, Arts and Letters* 6 (1885): 42–50.

Salzer, Robert J. "Other Late Woodland Developments." In *Introduction to Wisconsin Archaeology: Background for Cultural Resource Planning*, ed. William Green, James B. Stoltman, and Alice B. Kehoe. *Wisconsin Archeologist* 67, nos. 3, 4 (1986): 302–13.

———. "The Wisconsin North Lakes Project: A Preliminary Report." In *Aspects of Upper Great Lakes Prehistory: Papers in Honor of Lloyd A. Wilford*, ed. Elden Johnson, pp. 40–54. Minnesota Prehistoric Archaeology Series 11. St. Paul: Minnesota Historical Society, 1974.

Schoolcraft, Henry Rowe. *Narrative Journal of Travels from Detroit Northwest through the Great Chain of American Lakes to the Sources of the Mississippi River in the Year 1820*. New York: Arno and New York Times, 1970.

Schroeder, David L., and Katherine C. Ruhl. "Metallurgical Characteristics of North American Prehistoric Copper Work." *American Antiquity* 33, no. 2 (1968): 162–69.

Shay, C. Thomas. *The Itasca Bison Kill Site: An Ecological Analysis*. Minnesota Prehistoric Archaeology Series 6. St. Paul: Minnesota Historical Society, 1971.

Shimada, Izumi. "Pre-Hispanic Metallurgy and Mining in the Andes: Recent Advances and Future Tasks." In *In Quest of Mineral Wealth: Aboriginal and Colonial Mining and Metallurgy in Spanish America*, ed. Alan K. Craig and Robert C. West, pp. 37–73. Geoscience and Man 33. Baton Rouge: Louisiana State University Press, 1994.

Smith, Beverley Ann. "Systems of Subsistence and Networks of Exchange in the Terminal Woodland and Early Historic Periods in the Upper Great Lakes." Ph.D. diss., Department of Anthropology, Michigan State University, 1996.

Smith, Cyril. "Metallographic Study of Early Artifacts Made from Native Copper." *Actes du XIe Congrès t. VI* [Proceedings of the Eleventh Congress of the History of Science]. Warsaw, 1965. Pp. [237]–43.

Sommers, Lawrence M. *Atlas of Michigan*. East Lansing: Michigan State University Press, 1977.

Squier, Ephraim G., and Edwin H. Davis. *Ancient Monuments of the Mississippi Valley*. Smithsonian Contributions to Knowledge 1. Washington D.C.: Smithsonian Institution, 1848.

Steinbring, John Henry [Jack]. "Early Copper Artifacts in Western Manitoba." *Manitoba Archaeological Quarterly* 1, no. 1 (1991): 25–61.

———. "Old Copper Culture Artifacts in Manitoba." *American Antiquity* 31, no. 4 (1966): 567–74.

———. "The Preceramic Archaeology of Northern Minnesota." In *Aspects of Upper Great Lakes Anthropology: Papers in Honor of Lloyd A. Wilford*, ed. Elden Johnson, pp. 64–73. Minnesota Prehistoric Archaeology Series 11. St Paul: Minnesota Historical Society, 1974.

———. "Taxonomic and Associational Considerations of Copper Technology during the Archaic Tradition." Ph.D. diss., Department of Anthropology, University of Minnesota, 1975.

Stevens, John. "The Manufacture of Copper Beads Found in Michigan's Keweenaw Peninsula." Unpublished ms. on file at the Archaeology Laboratory, Michigan Technological University, Houghton, 1996.

Stoltman, James B. "The Archaic Tradition." In *Introduction to Wisconsin Archaeology: Background for Cultural Resource Planning,* ed. William Green, James B. Stoltman and Alice B. Kehoe. *Wisconsin Archeologist* 67, nos. 3, 4 (1986): 207–38.

———. *The Laurel Culture in Minnesota.* Minnesota Prehistoric Archaeology Series 8. St. Paul: Minnesota Historical Society, 1973.

Sutter, Lawrence. "Metallurgical Examination of Copper Artifacts from 20KE20." *Michigan Archaeologist* 39, nos. 3, 4 (1993): 166–70.

Thomas, Matthew M. "Preliminary Report of Investigations: Prehistoric and Historic Native Copper Mining in Northwestern Wisconsin." Unpublished paper, Department of Anthropology, University of Minnesota, Duluth, 1993.

Thwaites, Reuben G. *The French Regime in Wisconsin-I, 1637–1727.* Collections of the State Historical Society of Wisconsin 16. Madison: The Society, 1902.

Trevelyan, Amelia M. "Prehistoric Native American Copperwork from the Eastern United States." Ph.D. diss., Department of Art History, University of California, Los Angeles, 1987.

Trigger, Bruce G. *The Children of Aataentsic: A History of the Huron People to 1660.* Montreal: McGill-Queen's University Press, 1987.

Tylecote, R. F. *A History of Metallurgy.* 2d ed. London: Institute of Materials, 1992.

Varney, Tamara L., and Susan Pfeiffer. "The People of the Hind Site." *Ontario Archaeology* 59 (1995): 96–108.

Vecsey, Christopher. *Traditional Ojibwa Religion and Its Historical Changes.* Philadelphia: American Philosophical Society, 1983.

Vehik, Susan C., and Timothy G. Baugh. "Prehistoric Plains Trade." In *Prehistoric Exchange Systems in North America,* ed. Timothy G. Baugh and Jonathon E. Ericson, pp. 249–74. New York: Plenum, 1994.

Vernon, William W. "New Archaeometallurgical Perspectives on the Old Copper Industry of North America." In *Archaeological Geology of North America,* ed. N. Lasca and J. Donahue, pp. 499–512. Centennial Special Volume 4. Boulder: Geological Society of America, 1990.

Walthall, John A., et al. "Galena Analysis and Poverty Point Trade." *Midcontinental Journal of Archaeology* 7, no. 1 (1982): 133–48.

Warren, William W. *History of the Ojibway People.* St. Paul: Minnesota Historical Society Press, 1984.

Wayman, Michael L. "Native Copper: Humanity's Introduction to Metallurgy? Part 2: Metallurgical Characteristics and Utilization." *CIM, Canadian Institute of Mining and Metallurgy Bulletin* 78, no. 881 (1985): 75–77.

———. "Neutron Activation Analysis of Metals: A Case Study." *MASCA Research Papers in Science and Archaeology* 6 (1989): 66–71.

REFERENCES

———. "Recent Trends in Archaeometallurgical Research: Discussion." *MASCA Research Papers in Science and Archaeology* 8, no. 1 (1991): 79–82.

Wayman, Michael L., J. C. H. King, and P. T. Craddock. *Aspects of Early North American Metallurgy*. Occasional Paper 79. London: British Museum, 1992.

Wellman, Howard B. "The Provenance of Copper Artifacts from the Boucher Site." Master's thesis, Department of Archaeology, Boston University, 1993.

West, George A. "Copper: Its Mining and Use by the Aborigines of the Lake Superior Region." *Bulletin of the Public Museum of the City of Milwaukee* 10, no. 1 (1929): 1–182.

Whittlesey, Charles C. "Ancient Mining on the Shores of Lake Superior." *Smithsonian Contributions to Knowledge* 13, no. 4 (1863): 1–[32].

———. "Descriptions of Ancient Works in Ohio." *Smithsonian Contributions to Knowledge* 3, no. 7 (1850): 1–20.

Wilford, Lloyd A. "A Tentative Classification of the Prehistoric Cultures of Minnesota." *American Antiquity* 6, no. 3 (1941): 231–49.

Willey, Gordon R., and Jeremy A. Sabloff. *A History of American Archaeology*. 2d ed. San Francisco: W. H. Freeman, 1980.

Williams, Stephen. *Fantastic Archaeology: The Wild Side of North American Prehistory*. Philadelphia: University of Pennsylvania Press, 1991.

Willoughby, Charles C. "Michabo, the Great Hare: A Patron of the Hopewell Mound Settlement." *American Anthropologist* 37 (1935): 280–86.

———. "Primitive Metal Working." *American Anthropologist* 5 (1903): 55–57.

Wilson, Curtis, and Melville Sayre. "A Brief Metallographic Study of Primitive Copper." *American Antiquity* 2 (1935): 109–12.

Wilson, Daniel. The Ancient Miners of Lake Superior. *Canadian Journal* 3 (1856): 225–37.

Wilson, James Grant, ed. *Appleton's Cyclopedia of American Biography*. New York: D. Appleton and Company, 1888–1893.

Winchell, N. H. "Ancient Copper-mines of Isle Royale." *Popular Science Monthly* 19, no. 5 (1881): 601–20.

Winn, Vetal. "Ornamented Coppers of the Wisconsin Area." *Wisconsin Archeologist* 23, no. 3 (1942): 49–85.

Winters, Howard D. "Value Systems and Trade Cycles of the Late Archaic in the Midwest." In *New Perspectives in Archaeology*, ed. Sally R. Binford and Lewis R. Binford, pp. 175–221. Chicago: Aldine, 1968.

Wittry, Warren L. "A Preliminary Study of the Old Copper Complex." *Wisconsin Archeologist* 38, no. 4 (1957): 204–21.

———. "The Raddatz Rockshelter, SK 5, Wisconsin." *Wisconsin Archeologist* 40, no. 1 (1959): 33–69.

Wittry, Warren L., and Robert E. Ritzenthaler. "The Old Copper Complex: An Archaic Manifestation in Wisconsin." *Wisconsin Archeologist* 38, no. 4 (1957): 311–29.

Wood, Alvinus Brown. "The Ancient Copper-mines of Lake Superior." *Transactions of the American Institute of Mining Engineers* (containing the papers and discussions of 1906) 37 (1907): 288–96.

Wright, Gary A. "Some Aspects of Early and Mid-seventeenth Century Exchange Networks in the Western Great Lakes." *Michigan Archaeologist* 13, no. 4 (1967): 181–97.

Wright, James V. "An Archaeological Survey along the North Shore of Lake Superior." *Anthropology Papers of the National Museum of Canada* 3 (March 1963): 1–9.

———. *A History of the Native People of Canada.* Mercury Series Paper 152. Hull, Quebec: Canadian Museum of Civilization, 1995.

———. "The Prehistoric Transportation of Goods in the St. Lawrence River Basin." In *Prehistoric Exchange Systems in North America,* ed. Timothy G. Baugh and Jonathon E. Ericson, pp. 47–71. New York: Plenum, 1994.

Yarnell, Richard A. *Aboriginal Relationships between Culture and Plant Life in the Upper Great Lakes Region.* Anthropological Papers 23. Ann Arbor: Museum of Anthropology, University of Michigan, 1964.

Index

Accelerator mass spectrometry (AMS) dating, 143–45, 158, 159, 161, 162
Adena culture, 217, 241
Adz, adze. *See* Celts
Agassiz, Louis, 57
Agate Basin projectile, 146, 149, 150
Agricola, Georgius, 82, 94, 255n. 1 (top)
Agriculture, 168, 175–76
Algodonite, 134, 256n. 2 (top)
Algoma stage, 256n. 3 (bottom)
Ali Kosh, Iraq, 121
Alligator Eye site, 166–67
American Anthropologist, 64, 113
American Ethnological Society, 47
Analogy. *See* Arguments by analogy
Analytical methods: atomic absorption spectroscopy, 116, neutron activation analysis, 32, 116; optical emission spectroscopy, 32; radiocarbon dating, 143–44, 154–55, 256n. 1, 257n. 5 (top), 257n. 6; spectrographic analysis, 32; thin-sectioning of minerals, 61; X-ray fluorescence, 116
Andes Mountains, mining in, 106–7
Animal species, 36, 39–40, 149, 150, 152, 154, 155, 163, 166, 167, 169–70, 172, 173, 176, 178, 179, 191, 192, 195, 205–6
Animikig. *See* Manitou, Animikig
Annealing. *See* Copper, annealing
Anthropological Society of Washington, 113
Archaeological: context, 70, 219, 221, 242, 253n. 2, 256n. 4; culture(s), 141–42, 185, 191; deposits, characteristics of, 51, 139–40, 219; record, 15, 139; secondary deposits, 219; sites, 139, *map* 146, *map* 169
Archaeology: assumptions, 18; jargon, 18; public participation, 215, 220–24, 258n. 1 (top), 258n. 2; theory, 19
Archaeometallurgy, 114–16
Archaic Tradition, 119, 125, 130, 132, 142, *map* 146, 152–68, 185, 189–90, 192; Early Archaic stage, 142, 145, 150, 151, 153–54; Late Archaic stage, 122, 124, 127, 129, 142, 148, 162–67, 186, 187, 193, 198; Middle Archaic stage, 142, 154–62, 186, 192, 194
Archean Eon, 26
Arctic mirage. *See* Mirage effect
Arguments by analogy, 47, 52
Arguments of association, 49, 50–51, 70
Artifacts, 227–52, 258n. 1 (bottom); care of, 256n. 2 (bottom); collections and collecting, 18, 220–23, 224, *illus.* 228, 256n. 2 (bottom), 256n. 4, 257n. 9, 259n. 2; distribution, 185–92, 199, 257n. 2; documentation

277

of, 18; donation of, 259n. 2; effigies, 170, 210, 244; function, 49, 170, 174, 227–52; replication of, 113–15, 118–23, 125–26, 232–33; ritual beliefs, 204–9; site destruction, 220–23, 253n. 2, 257n. 4 (top). *See also* Awls; Beads; Celts; Crescent knives; Fishhooks; Knives; Ornaments; Projectile points; Spatulas; Spuds
Asia, metal-working, 117
atl atl, 152, 248
Atomic absorption spectroscopy, 116
Aurora borealis. *See* Northern lights
Awls, 49, 156, 157, 158, 162, 167, 170, 171, 172, 173–74, 175, 180, 187, 188, 229–33, *illus.* 230–31
Ax, Axe. *See* Celts
Aztalan, 139

Barrel copper, 55, 89
Bastian, Tyler, 76–78
Beads, *illus.* 192, 233–35, *illus.* 234. *See also* Ornaments
Beaubien, Paul L., 74
Blackduck, 142, 241
Boucher site, 132, 133, 190–92, 235
Bracelet. *See* Ornaments
Brewerton phase, 186
British Museum, 45, 135, 254n. 6
Bronze Age, 72, 217–18; in Europe, 20; misidentification of copper as bronze, 133–36, 218
Brule River, 28
Burials. *See* Mortuary sites
Burnt Rollways phase, 167

Cache deposit, 139, 173, 188
Caledonia, 85–87, 97, 98–105, 255n. 5
Canadian Biome, 39
Canadian Shield, 26, 141, 153, 171, 175, 210
Catlinite, 194
Celts, 157, 160, 162, 170, 172, 175, 187, 188, 229, 233, 235–38, *illus.* 236–37, 252

Ceramics, 168–80, 257n. 7; earliest, 168, 257n. 7; Dane Incised, 168; Laurel Ware, 168, 170, 172, 174, 241; trade and exchange, 174–75, 176, 178
Chautauqua Grounds site, 155, 161, 171, 238, 249
Chequamegon, 204
Chert, 153, 165, 171, 174, 175, 186, 194, 195; Bayport formation, 165; Burlington formation, 195; Hudson Bay lowland materials, 165; Knife River formation, 195; ritual beliefs, 206–7
Chippewa people. *See* Native American people, Ojibwe
Chisel. *See* Celts
Chynoweth, Benjamin F., 255n. 5
Clark, Caven P., 76
Cliff Mine, 28, 107
Climate, 40–41, 154
Collecting. *See* Artifact collections and collecting
Columbian Exposition (1892–1893), 64
Comparative method, 54, 65–66
Conglomerate bedrock, 27
Coolidge, Calvin, 69, 254n. 10
Copper: alloys, 30, 116, 131–37, 256n. 2 (top); amygdaloid deposits, 77; annealing, 115–16, 120–23, 170, 177; bonding, 116, 129, 131–37; casting, 116, 128; chemical sourcing of, 30–32; cladding, 127–28; cold working 116–17, 137; craft specialists, 119; cutting, 116, 125–26; deformation, 117; discovery of, 69; distinguishing European 31, 32; drawing, 131; drift, 25, 30, 33–34, 50, 58, 83–84, 91, 113, 118, 150, 160, 179, 255n. 2; ductility, 117, 119–23; elemental composition, 30–32, 116, 192; embossing, 116, 125–26; flow lines, 117; folding, 116, 129, 121, 177; genesis of, 26–28; geological occurrences, *illus.* 24, 25, 26–28, 31; glacial dispersal

INDEX

of, 25, 33–34, 30, 150, 255n. 3; grain structure, 117, 119–23; grinding, 116, 124–25; hammering, 116, 118–19, 177; hardening, 54, 114–15, 119–23, 215; heat treatment, 114, 120–23, 131–37, 256n. 3 (top); hot forging, 116, 131–37, 224; joining, 116, 132–33; lodes, 60; mandrel, use of, 127; melting, 59; melting point, 59, 116, 135; microstructure, 114–17, 122, *illus.* 192; molding, 114, 127–28; native deposits, 26–28, 31; perforating, 116, 125–26; polishing, 116, 126–27, 130; preservation of organics, 143, 161, 173, 224, 256n. 2 (bottom); pressure welding, 116, 133; punching, 126; purity, 30; quantity mined, 55, 56, 70, 71, 73; recrystallization, 115, 117, 119–23; riveting, 116, 126, 129–30; rolling, 129; silver inclusions, 27, 30, 50; sinking, 116, 127–28; smelting evidence, 29, 30, 64, 116, 221, 253n. 2; sources, 31, 78, 119, 190, 192, 212, 224, 253n. 2; stone working, analogies with, 117, 216, 255n. 1 (bottom); stretch hammering, 116, 118–19; supernatural beliefs, 70, 163, 183–84, 191, 199–204, 208–9, 216, 255n. 1 (top), 258n. 7; swedging, 116, 118, 130; trace elements in, 16, 30–32, 116; trade and control, 70, 71–72, 177, 187–88, 191–99, 216, 224; variability of deposits, 28–30, 55, 89; yield points, 121, 233
Copper Age, 19, 63, 72
Copper Falls, 107
Copper Harbor, 28, 84
Cranbrook Institute of Science, 259n. 2
Cremation. *See* Mortuary sites
Crescent knives, 157, 158, 160, 162, 165, 173–74, 238–42, *illus.* 239–40
Cult archaeology, 216–20
Cultural evolution, views on, 46, 63, 72–73

Cummins site, 151
Curated assemblage, 105, 167
Cushing, Frank H., 113–14, 116–19, 125, 131

Dane Incised ceramics, 168
Davis, Edwin H., 31, 46–51, 56, 122, 127, 133–36, 253n. 2, 254n. 6
Deer Lake sites, 147, 149, 164, 165–66
Delaware, 208
Delaware Mine, 107
Dendrochronology, 48, 53, 60
Drier, Roy W., 71–74, *illus.* 72
Drill. *See* Awls
Dumaw Creek site, 180, 244
Dustin, Fred, 75
Du Temple, Octave, 71–74

Earspools. *See* Ornaments
Eden-Scottsbluff projectiles, 146, 148–49
Effigy mounds, 48
Etowah site, 31, 123, 128, 129
Europe: copper-working technology and smithing, 114, 117; entrada to North America, 48, 53, 54, 199; introduction of trade materials, 30, 63, 180–81; mining technology, 82–83; trading in copper, 20, 71
Excavation of prehistoric mining pits, 60, 64–65, 74, 89
Exotic materials. *See* Trade, goods
Expedient technology, 105, 166–67
Experimental archaeology, 49; in copper mining, 55, 77, *illus.* 92, 92–93, 95; in metal-working, 16, 113–15, 118–23, 125–26, 133, 224, 256n. 3 (top). *See also* Artifacts, replication of

Feast of the Dead. *See* Mortuary sites
Fell, Barry, 217–19
Field Museum of Natural History (Chicago), 232, 244, 248
Fire setting and mining. *See* Mining, fire setting and charcoal

Fishhooks, 158, 167, 172, 175, 180, 188, *illus.* 230, 242
Fishing technologies, 153, 155, 161, 170, 172, 176, 179, 188
Fissure mines and veins, 30, 56, 77, 89, 90–92, 95
Flambeau phase, 142, 149
Float copper. *See* Copper, glacial dispersal of
Fogel, Ira L., 185–87
Forest Mine, 88, 108
Formal tools. *See* Curated assemblage
Foster, John W., 36, 38, 41, 51–53, *illus.* 52
Fox, George R., 95

Gad. *See* Celts
Gaff, 242
Galena, 186, 189, 194–95
Gannon, Joseph C., 73
Garden beds, prehistoric, 48
Geology: bedrock, 26; Lake Superior basin, 23, *illus.* 25, 26–28
Glacial geology, 57; and copper, 25, 30, 33–34, 58, 63, 68, 83–84, 148, 150, 160, 255n. 2, 255n. 3; drift deposits, 195; early theories, 57; post-glacial environments, 36, 145–46; post-glacial Great Lakes, 34, 150–51, 154, 256n. 3 (bottom); sequence, 33–36; surface formations, 33–36. *See also* Pleistocene Epoch
Glacial Kame Complex, 142, 153, 160, 186, 189–90
Gorge. *See* Fishhooks
Gorget. *See* Ornaments
Gouge. *See* Celts
Greenstone lava flow, 27
Gribben forest 33, 34
Griffin, James B., 74–76, *illus.* 75
Gros Cap site, 180–81
Guthe, Carl, 75

Hafts: artifact, 18, 91, 98–107, 233, 238–41, 244; experimental, 102; materials, 98, 249; polish, 65, 102; riveting, 129–30; shimming, 102; sockets, 125, 252; wood and organic residues, 106–7, 143, 145, 224, 249, 256n. 2 (bottom)
Hammerstones: alteration, 73, 98–104; comparisons, 52, 59, 96–107; durability, 92, 97; experimental use, 92; hafting, 65–66, 91, 105–7; hardness, 105; Isle Royale, 65, *illus.* 87, 96–98, *illus.* 100–101, *illus.* 103, 105; mainland, 86, 96, 105; numbers, 76; polish, 65, 102; rock selection, 70, 97, 104–5; sources, 73; use-wear, 97, 104; variability, 97–98, *illus.* 103
Harpoon. *See* Projectile points
Harrison, William Henry, 254n. 4 (bottom)
Harz Mountains, Germany, 52
Hematite. *See* Ochre
Henry, Alexander, 54
Henry, Joseph, 47
Heterarchy, 257n. 5 (bottom)
Hillingar effect. *See* Mirage effect
Hind site, 145, 189–90
Historic period, 142, 179, 208–9
Hixton orthoquartzite, 147–49, 151, 153, 165, 166, 193
Holmes, William H., 15, 64–66, *illus.* 65
Hopewell culture, 122, 125, 168, 174, 175, 194, 241, 258n. 1 (bottom); site, 113–14, 116, 118, 122, 126–28, 129, 132, 133–36, 157, 172, 217, 218, 258n. 8
Hornstone, 162, 163, 194
Houghton County, Michigan, 252, 259n. 2
Houses and structures, prehistoric, 169–72, 175, 177
Hoy, P. R., 118, 130
Huantajaya, Peru, 106
Hunter's Point site, 207–8
Hunting magic, 213, 258n. 5
Huron people. *See* Native American people, Huron

Hydrothermal deposition of copper, 27–28
Hypothesis formulation and testing, 47, 62

Iconography, 184, 204–9; copper, 188; motifs, 206, 208–9, *illus.* 209, 235, 241, 244, 245, 252, 257n. 9
Ideology, 192; concerning copper, 199–204, 208–9; in mortuary ceremony, 155, 162–63, 191; offerings, 205–8, 211
Illinois, 33, 35, 150, 168, 176, 187, 194, 255n. 3
Indiana, 168, 255n. 3
Ingot, copper, 257n. 8
Interlakes composite, 142, 144, 150–51
Iowa, 255n. 3
Iron ore, 158, 160
Iroquois people. *See* Native American people, Iroquois
Isle Royale, 26, 27, 28, *illus.* 38, 52, 56, 57, 64, 66–71, 83, 85, 88, 89, 94, 95, 98–105, 108, 153, 174, 178–79, 193, 194, 202; description, 37, 38–39, 43, 254n. 4 (top); duration of mining, 160, 163–64; McGargoe's Cove, 68, 107; Minong Ridge, 28, 68, 75, 83, 90, 95, 155; National Park, 69, 73; Pit 25, 75; Pit 54, 95; "pit houses," 67–68; prehistoric research, 73–78, 255n. 4; Siskowit Lake, 194
Itasca site, 149, 150, 154

Jasper taconite, 153, 193
Juntunen site, 132–33, 142, 176–78

Keweenaw Peninsula, Michigan, 26, 27, 28, 33, 35, 36, 40–42, 52, 55, 60, 71, 76, 88, 107, 143, 160, 238, 242, 256n. 3 (bottom)
Knapp, Samuel O., 52, 60, 83, 87
Knives, 157, 159, 160, 180, 187, 188, 208, 242–44, *illus.* 243

Lake Huron, 140, 195, 207; glacial history, 34

Lake Michigan, 140; glacial history, 34
Lake Nipigon, Ontario, 188
Lake Superior, 15, 140–41, 187; archaeological sites, 140, 143–48, 150, 152–67, 242, 244; climate, 40–41, 254n. 4 (top); description, 23; distribution of copper in, 26; geology of 26–28; glacial history, 33–36, 150–51, 256n. 3 (bottom); landscape, 36–43; prior research about, 64; sacred meaning, 40; significance to archaeology, 49; storms, 41; syncline, 27
Lakes phase, 142, 179, 180
Landscape: cultural production of, 23, 40; historical description, 36–42
La Roche Verte, 84
Late Woodland period, 133. *See also* Woodland Tradition, Terminal Woodland stage
Lava, 26–28
Law of Superposition, 49, 254n. 5
Lead. *See* Galena
Lithics trade, 193–95. *See also* Chert; Hixton orthoquartzite; Hornstone; Jasper taconite; Obsidian; Quartz; Quartzite
Lode mine, 90–92
Lookout site, 155

Mackinac phase, 142
Magma, 26
Manitoba, 150, 178; Rainy River district, 168
Manitou, 42, 58, 162, 184, 199–204, 206, 210–13, 257n. 4 (bottom), 257n. 5 (bottom); and copper, 16; Animikig, 200–201, 203, 204, 205; behaviors, 199–201; manitoussiwuk, 203; Mishi Bizi, 42, 201–2, 203, 206–7, 210; Mishi Ginabig, 42, 202–3, 205–6, 258n. 8; mizauwabeekummoowuk, 203; mutchimanitou, 202; Nanabohzo, 42, 201, 203, 204; sky, 200–201,

203, 204, 205, 207; underwater, 200–201, 205, 210
Manitoulin Island, Ontario, 195
Marquette County, Michigan, 33, 147
Mass City, Michigan, 28, 97
Mass copper, 30, 60, 89
Massee, Burt A., 66–71, 194, 254n. 10
Mass Mine, 255n. 5
McCreary point, 150
McDonald, Eugene F., Jr., 66–71, 194, 254n. 10
McPherron, Alan, 176–78
Meadowood phase, 191, 257n. 1
Menominee people. *See* Native American people, Menominee
Menominee River, 37, 162
Metal detecting: collaboration with archaeology, 221–23; destructive impacts of, 20, 220–21; United Kingdom, 220–21; United States, 222
Metallography, 114–16, 137
Metallurgy, 256n. 2 (top); universal development of, 63
Metal production system, 81
Mexico: metal working, 135; mines, 106
Michigan, 35, 48, 139, 145–46, 150, 155, 156, 195, 208, 257n. 4 (top)
Michigan Archaeological Society, 223, 258n. 1 (top)
Michigan College of Mining and Technology. *See* Michigan Technological University
Michigan State Geological Survey, 97, 255n. 5
Michigan Technological University, 71, 74, 75, 97, 254n. 11, 257n. 9; Archaeology Laboratory, 98; Seaman Mineralogical Museum, 97
Michipicoten Island, Ontario, 203
Midden deposit, 170, 171, 179
Middlesex phase, 191, 257n. 1
Middle Woodland period, 78, 119, 123, 125, 127–28, 157, 194. *See also* Woodland Tradition, Initial Woodland stage
Midewiwin Society, 208
Migration theories, 51, 56
Milwaukee Public Museum, 66–67, 68, 157, 158, 162, 254n. 10
Minerals in association with copper, 27, 69
Miners, prehistoric: "inscrutability," 73, 84; numbers, 72; prospecting success, 59–60, 82; question of ethnic identity, 52, 56, 58, 60–61, 62–63, 65, 67, 70–72, 216, 219–20; work habits, 54, 56, 68–70, 73
Minesota Mine, 52, 60, 61, 87–88, 89, 95, 108, 254n. 7
Mining: antiquity of, 53, 58–59, 60–61, 71; description, 60, 64, 84–96; European, 255n. 1 (top); evidence, *illus.* 28, 84–91; excavations, experimental, 77, *illus.* 92, 92–93, 95; fire setting and charcoal, 60, 68, 77, 90, 93–96, 110, 166–67; gutters, 77, 91, 109; labor requirements, 69–70, 72, 77, 95–96; ladders, 108; management and social control, 69, 110–11; pits, archaeological excavation, 60, 64–65, 74, 89, 95; pits, extent and number, 59, 76, 77, 84–91; pits, variability and appearance, 58, 77, 84–91, *illus.* 85, *illus.* 88; prospection, 82–83; quarrying, 64, 91; quenching, 93–96; scale of productivity, 63, 73; scars on bedrock, 58–59, 89, 94; sequence, 91–93; skids, 109; stamping, 76, 93; tailings, 90–91, 92; technology, 91–96; timbering and scaffolding, 108–9; tools, 96–110; tunneling, lack of, 90; water washing, 59, 255n. 2
Minnesota, 148, 149, 150, 154, 194, 195, 207–8, 210
Minnesota Historical Society, 232
Minong Mine, 95, 155
Minoqua phase, 142, 149

Mirage effect, 41–42, 43, 201
Mishi Bizi. *See* Manitou, Mishi Bizi
Mishi Ginabig. *See* Manitou, Mishi Ginabig
Missionaries, 58; Allouez, 201, 202, 203, 204, 258n. 6; Dablon, 64, 203
Mississippi River headwaters, 195
Mississippian cultures, 122, 125, 126, 128, 129, 258n. 1 (bottom)
Moorehead, Warren K., 113
Morrison's Island-6 site, 188, 208, 232
Mortuary sites, 153, 155, 157–63, 167–68, 177–78, 179–80, 185–86, *map* 186, 187, 188–91; children in, 155, 159, 162–63, 189, 190, 210; cremation, 148, 151, 157, 158, 159, 186, 194, 196; women in, 155, 162–63, 177–78, 180, 189, 190–91
Moundbuilders and mounds, 16, 48, 50, 168, 175, 177, 179, 186, 217; antiquity, 48–49, 50; ethnic identity, 56; links with copper miners, 46, 50–51, 54, 58, 61–62, 113, 118, 135, 253n. 2, 254n. 6; pottery, 253n. 2
Mt. Gabriel, Ireland, 93, 95–96, 98, 106
Mythology 82, 199–204, 216–20

Nanabohzo. *See* Manitou, Nanabohzo
Naomikong Point site, 172, 194
National Mine, 255n. 5
Native American people: Algonquian language family, 70–71, 217, 257n. 3; group interaction, 187–89, 210–13; Huron, 142, 213; Iroquois, 142, 177, 178, 199, 210; Menominee, 71, 180, 202; Odawa (Ottawa), 142, 195; Ojibwe (Ojibwa, Chippewa), 70–71, 142, 201, 258n. 5; Pima, 217; population, 170, 178; trade relations, 195–99, 210–13; Siouan language family, 70; Sioux, 70–71; Susquehannock, 210, 213; Zuni, 113–14, 131, 217
Nebraska, 255n. 3
Needle. *See* Awls
Neutron activation analysis, 32, 116

New York, 176, 187, 188, 195, 208, 257n. 1
New York Ethnological Society. *See* American Ethnological Society
Nipissing stage, 34, 154, 256n. 3 (bottom)
Nokomis phase, 174–75, 194
Northern lights, 23, 41
North Lakes sites, 148–49, 153–54, 164, 167, 174–75, 194, 195
North West Mine, 56, 84, 89
Norton Mounds, 139
Nova Scotia, 192

Obsidian, 162, 163, 172, 175, 194; sources, 51, 194
Ochre, 153, 159, 162, 185–86, 189, 206. *See also* Red Ochre Complex
Oconto County Historical Society, 158
Oconto site, 144–45, 155, 158–59, 160, 161, 196
Odawa people. *See* Native American people, Odawa
Ohio, 56, 168, 192, 194, 255n. 3
Ojibwa people. *See* Native American people, Ojibwe
Old Copper Complex, 122, 124, 142, 153, 156–61, 163, 185, 190, 257n. 4 (top), 257n. 6; core artifacts, 156, 165, 173–74, 191–92, 241–42
Old World: contact with, 20; mythology of trade with, 216–20
Oneota culture, 142
Ontario, 145, 148, 176, 178, 180, 188, 189, 213, 242
Ontonagon Boulder, *illus.* 29, 37
Ontonagon County, Michigan, 28, 35, 52, 54, 60, 66, 76, 191, 255n. 5
Ontonagon River, 35, 36, 37, 39, 41, 83, 211, 224, 254n. 3 (top), 254n. 12, 255n. 6
Optical emission spectroscopy, 32
Organic preservation, 159, 162, 195–96, 224
Ornaments, 49, 116, 118, 121, 122–23, 127–28, *illus.* 129, 131, 133, 157,

160, 162, 170–71, 172, 173–74, 175, 179, 180, 186, 187, 188, 190, 192, 207–8, 233–35, *illus.* 234, 241, 242, 244, *illus.* 245
Osceola site, 157–58, 159–60, 194–95, 232, 252
Oshkosh Public Museum, 159, 162
Ossuary. *See* Mortuary sites
Ottawa National Forest, 166, 171
Ottawa North site, 166–67
Ottawa people. *See* Native American people, Odawa
Ottawa River, 188

Packard, R. L., 63–64
Paleoindian Tradition, 142, 145–52, map 146, 156, 255n. 1 (bottom). *See also* Plano industries
Papworth, Mark, 75
Perforator. *See* Awls
Phoenix Mine, 108
Pic River site, 180
Pick. *See* Awls
Pike. *See* Awls
Pin. *See* Awls
Pipestone. *See* Catlinite
Plains area, 194, 195
Plano industries, 148, 150, 165
Plant species, 36; agriculture, 168, 175–76, 179; use by native peoples, 39, 163, 167, 170, 171, 173, 195, 249. *See also* Vegetation
Pleistocene Epoch, 26, 33, 146
Point Peninsula culture, 186
Porcupine Mountains, 37
Portage Lake, 28, 36, 57, 59, 60, 83
Portage Lake Ship Canal, 55, 63
Pottery. *See* Ceramics
Poverty Point, Louisiana, 122
Private collections. *See* Artifact collections and collecting
Projectile points, 143, 146, 147, 148, 149, 150–51, 154, 157, 158, 159, 160, 161, 162, 175, 179, 180, 187, 188, 208, 244–49, *illus.* 246–47; manufacturing sequence, 248

Property analysis, 81
Prospecting methods, copper, 82–83
Proterozooic Eon, 26
Public involvement in archaeology, 221–24, 258n. 1 (top), 258n. 2
Punch. *See* Awls

Quartz, 165, 166, 171, 206
Quartzite, 153, 164, 165, 174
Quebec, 188
Quenching. *See* Mining, quenching
Quimby, George I., 75
Quincy Mine, ancient pit, 55, 89

Radiocarbon dating, 32, 74, 143–45, 154–55, 158; Isle Royale, 74, 76; statistical methods of, 26, 256n. 1, 257n. 5 (top), 257n. 6
Red Ochre Complex, 142, 153, 163, 185–86, 194
Reeder, John T., 257n. 9, 259n. 2
Reigh site, 159–60, 196, 241, 244, 249, 252
Renshaw site, 144, 249
Richter site, 175
Rift, 27
Riverside site, 161–63, 166, 173, 180, 190, 194, 195, 196, 233, 241, 249
Robinson site, 175
Rock art, 206–7
Rockland, 97
Rock shelters, 155
Rudna Glava mines, 106

Saint Lawrence River, 188
Sand Point site, 179–80
Saskatchewan, 150
Sault Sainte Marie, 41
Schoolcraft, Henry Rowe, 37, 41, 254n. 3 (top)
Science and the scientific method: assumptions, 46–47; challenges to, 215–20; characteristics of, 18; disproof, 46; hypothesis testing, 62; parsimony, 53, 68; skepticism, 20
Selkirk culture, 142

Semi-lunar knife. *See* Crescent knives
Shaw, C. G., 95
Sheet copper, 56
Shell, 153, 207; appearance in mortuary ceremony, 158, 160, 162, 163, 186, 189, 190–91, 193; trade, 162, 192, 195, 196
Shield Archaic, 142, 164
Shingebiss, 201
Silver, occurrence with copper, 27, 30, 50
Silver Mound, Wisconsin, 147
Siskowit Lake. *See* Isle Royale, Siskowit Lake
Site destruction, 15, 20; history, 48, 253n. 2, 257n. 4 (top). *See also* Threats to archaeological sites
Smithson, James. *See* Smithsonian Institution
Smithsonian Institution, 47, 56, 57, 63, 253n. 2
Society for American Archaeology, 258n. 2
Socketed tools, 125, 127, 157, 160, 165
South America, artifact collections, 136
South Fowl Lake site, 143–45, 149, 151
Spain, mines, 106
Spatulas, 159, 249–52, *illus.* 250–51
Spaulding, Albert, 75
Spears. *See* Projectile points
Spectrographic analysis, 32
Spuds, 157, 250–52, *illus.* 250–51
Squier, Ephraim G., 31, 46–51, 56, 122, 127, 133–36, 253n. 2, 254n. 6
Squirrel River phase, 142, 149, 153–54
State Historical Society of Wisconsin, 118, 145, 158, 159, 232
Straits of Mackinac, 175, 176–78, 194
Subduction zone, 26
Summer Island site, 169–71, 172, 238, 244

Territory, band or group, 105, 255n. 6
Textiles, iconography, 205; preservation of, 162, 173, 191
Thames River, 188

Theory, archaeological, 19
Thin-sectioning of materials, 61
Threats to archaeological sites: collecting, 220–23, 253n. 2, metal detecting, 220–23; shoreline development, 15
Thunderbird. *See* Manitou, Animikig
Time-space systematics, 48, 254n. 2
Timid Mink site, 171–72, 175
Tobacco, 195, 201, 206, 213, 258n. 6
Trace element analysis, 16, 25, 116, 192
Trade: directional trading model, 198; down-the-line exchange model, 174, 198–99; eastern North America, 16, 190–91, 195–96, 223–24; goods, 157, 162, 192–99; mechanisms, 163, 192, 198–99; motivations, 192–204, 210–13; perishable goods, 195–96; prehistoric, 25, 168, 173, 174–75, 177, 180, 184, 191–99, 210–13; routes, 163, 187–88, 193–94, 197–99; trans-Atlantic myth, 216–20
Trap range, 51, 255n. 6; definition, 253n. 1; description, 23, 26, 35–36; landscape, 36–37
Treaty of LaPointe, 64
Trent Waterway, 188, 192
20KE20, 143–44, 147, 164–65, 173–74, 192, 232, 241

Ulu. *See* Crescent knives
University of Michigan, 74, 176; and Isle Royale National Park, 74–78; Museum of Anthropology, 162, 172
University of Minnesota-Duluth Archaeometry facility, 32
University of Wisconsin, 159
Upper Peninsula, Michigan, 145, 157

Vegetation: description, 36; uses by aboriginal cultures, 39
Vermont, 190–91
Vesicles, copper bearing, 27
Violence, evidence of, 162, 180
Vision quest, 208

Wampum. *See* Shell
Waterbury Mine, 89, 107–9
Water panther. *See* Manitou, Mishi Bizi
Water transportation: and distribution of copper, 187–88; and trade, 188, 193–94
Wedge. *See* Celts
West, George A., 66–71, *illus.* 67, 254n. 10
West, Roy O., 69, 254n. 10
Whitney, Josiah D., 36, 38, 41, 51–53, *illus.* 52
Whittlesey, Charles C., 56–61, *illus.* 57
Willoughby, Charles C., 116–21, 125–26, 127, 256n. 2 (top)
Wilson, Daniel, 53–56, *illus.* 55
Winchell, N. H., 61–63, *illus.* 62
Winter site, 194
Wisconsin, 28, 35, 36, 48, 63, 71, 114, 118, 129, 139, 145, 146, 147–49, 155, 156–61, 185, 187, 188, 189, 191, 195, 196, 210, 241, 257n. 6
Wisconsin Archeological Society, 223, 258n. 1 (top)
Wisconsin Archeological Survey, 159
Wisconsin Archeologist, 160
Wittry, Warren L., 227, 229
Women: in mortuary sites, 177–78, 180, 189; tools and activities, 173–74
Wood, Alvinus B., 53–56
Woodland Tradition, 142, 157–58, 162, 168–80, *map* 169, 190, 199; community organization, 169–72; Early Woodland stage, 160, 257n. 1, 257n. 7; houses and structures, 169–72, 175, 177; Initial Woodland stage, 142, 157, 168–75, 194, 195, 207, 257n. 7; Terminal Woodland stage, 142, 148, 175–80

X-ray fluorescence, 116

Yellowstone, Wyoming, 194

TITLES IN THE GREAT LAKES BOOKS SERIES

Freshwater Fury: Yarns and Reminiscences of the Greatest Storm in Inland Navigation, by Frank Barcus, 1986 (reprint)

Call It North Country: The Story of Upper Michigan, by John Bartlow Martin, 1986 (reprint)

The Land of the Crooked Tree, by U. P. Hedrick, 1986 (reprint)

Michigan Place Names, by Walter Romig, 1986 (reprint)

Luke Karamazov, by Conrad Hilberry, 1987

The Late, Great Lakes: An Environmental History, by William Ashworth, 1987 (reprint)

Great Pages of Michigan History from the Detroit Free Press, 1987

Waiting for the Morning Train: An American Boyhood, by Bruce Catton, 1987 (reprint)

Michigan Voices: Our State's History in the Words of the People Who Lived It, compiled and edited by Joe Grimm, 1987

Danny and the Boys, Being Some Legends of Hungry Hollow, by Robert Traver, 1987 (reprint)

Hanging On, or How to Get through a Depression and Enjoy Life, by Edmund G. Love, 1987 (reprint)

The Situation in Flushing, by Edmund G. Love, 1987 (reprint)

A Small Bequest, by Edmund G. Love, 1987 (reprint)

The Saginaw Paul Bunyan, by James Stevens, 1987 (reprint)

The Ambassador Bridge: A Monument to Progress, by Philip P. Mason, 1988

Let the Drum Beat: A History of the Detroit Light Guard, by Stanley D. Solvick, 1988

An Afternoon in Waterloo Park, by Gerald Dumas, 1988 (reprint)

Contemporary Michigan Poetry: Poems from the Third Coast, edited by Michael Delp, Conrad Hilberry and Herbert Scott, 1988

Over the Graves of Horses, by Michael Delp, 1988

Wolf in Sheep's Clothing: The Search for a Child Killer, by Tommy McIntyre, 1988

Copper-Toed Boots, by Marguerite de Angeli, 1989 (reprint)

Detroit Images: Photographs of the Renaissance City, edited by John J. Bukowczyk and Douglas Aikenhead, with Peter Slavcheff, 1989

Hangdog Reef: Poems Sailing the Great Lakes, by Stephen Tudor, 1989

Detroit: City of Race and Class Violence, revised edition, by B. J. Widick, 1989

Deep Woods Frontier: A History of Logging in Northern Michigan, by Theodore J. Karamanski, 1989

Orvie, The Dictator of Dearborn, by David L. Good, 1989

Seasons of Grace: A History of the Catholic Archdiocese of Detroit, by Leslie Woodcock Tentler, 1990

The Pottery of John Foster: Form and Meaning, by Gordon and Elizabeth Orear, 1990

The Diary of Bishop Frederic Baraga: First Bishop of Marquette, Michigan, edited by Regis M. Walling and Rev. N. Daniel Rupp, 1990

Walnut Pickles and Watermelon Cake: A Century of Michigan Cooking, by Larry B. Massie and Priscilla Massie, 1990

The Making of Michigan, 1820–1860: A Pioneer Anthology, edited by Justin L. Kestenbaum, 1990

America's Favorite Homes: A Guide to Popular Early Twentieth-Century Homes, by Robert Schweitzer and Michael W. R. Davis, 1990

Beyond the Model T: The Other Ventures of Henry Ford, by Ford R. Bryan, 1990

Life after the Line, by Josie Kearns, 1990

Michigan Lumbertowns: Lumbermen and Laborers in Saginaw, Bay City, and Muskegon, 1870–1905, by Jeremy W. Kilar, 1990

Detroit Kids Catalog: The Hometown Tourist by Ellyce Field, 1990

Waiting for the News, by Leo Litwak, 1990 (reprint)

Detroit Perspectives, edited by Wilma Wood Henrickson, 1991

Life on the Great Lakes: A Wheelsman's Story, by Fred W. Dutton, edited by William Donohue Ellis, 1991

Copper Country Journal: The Diary of Schoolmaster Henry Hobart, 1863–1864, by Henry Hobart, edited by Philip P. Mason, 1991

John Jacob Astor: Business and Finance in the Early Republic, by John Denis Haeger, 1991

Survival and Regeneration: Detroit's American Indian Community, by Edmund J. Danziger, Jr., 1991

Steamboats and Sailors of the Great Lakes, by Mark L. Thompson, 1991

Cobb Would Have Caught It: The Golden Age of Baseball in Detroit, by Richard Bak, 1991

Michigan in Literature, by Clarence Andrews, 1992

Under the Influence of Water: Poems, Essays, and Stories, by Michael Delp, 1992

The Country Kitchen, by Della T. Lutes, 1992 (reprint)

The Making of a Mining District: Keweenaw Native Copper 1500–1870, by David J. Krause, 1992

Kids Catalog of Michigan Adventures, by Ellyce Field, 1993

Henry's Lieutenants, by Ford R. Bryan, 1993

Historic Highway Bridges of Michigan, by Charles K. Hyde, 1993

Lake Erie and Lake St. Clair Handbook, by Stanley J. Bolsenga and Charles E. Herndendorf, 1993

Queen of the Lakes, by Mark Thompson, 1994

Iron Fleet: The Great Lakes in World War II, by George J. Joachim, 1994

Turkey Stearnes and the Detroit Stars: The Negro Leagues in Detroit, 1919–1933, by Richard Bak, 1994

Pontiac and the Indian Uprising, by Howard H. Peckham, 1994 (reprint)

Charting the Inland Seas: A History of the U.S. Lake Survey, by Arthur M. Woodford, 1994 (reprint)

Ojibwa Narratives of Charles and Charlotte Kawbawgam and Jacques LePique, 1893–1895. Recorded with Notes by Homer H. Kidder, edited by Arthur P. Bourgeois, 1994, co-published with the Marquette County Historical Society

Strangers and Sojourners: A History of Michigan's Keweenaw Peninsula, by Arthur W. Thurner, 1994

Win Some, Lose Some: G. Mennen Williams and the New Democrats, by Helen Washburn Berthelot, 1995

Sarkis, by Gordon and Elizabeth Orear, 1995

The Northern Lights: Lighthouses of the Upper Great Lakes, by Charles K. Hyde, 1995 (reprint)

Kids Catalog of Michigan Adventures, second edition, by Ellyce Field, 1995

Rumrunning and the Roaring Twenties: Prohibition on the Michigan-Ontario Waterway, by Philip P. Mason, 1995

In the Wilderness with the Red Indians, by E. R. Baierlein, translated by Anita Z. Boldt, edited by Harold W. Moll, 1996

Elmwood Endures: History of a Detroit Cemetery, by Michael Franck, 1996

Master of Precision: Henry M. Leland, by Mrs. Wilfred C. Leland with Minnie Dubbs Millbrook, 1996 (reprint)

Haul-Out: New and Selected Poems, by Stephen Tudor, 1996

Kids Catalog of Michigan Adventures, third edition, by Ellyce Field, 1997

Beyond the Model T: The Other Ventures of Henry Ford, revised edition, by Ford R. Bryan, 1997

Young Henry Ford: A Picture History of the First Forty Years, by Sidney Olson, 1997 (reprint)

The Coast of Nowhere: Meditations on Rivers, Lakes and Streams, by Michael Delp, 1997

From Saginaw Valley to Tin Pan Alley: Saginaw's Contribution to American Popular Music, 1890–1955, by R. Grant Smith, 1998

The Long Winter Ends, by Newton G. Thomas, 1998 (reprint)

Bridging the River of Hatred: The Pioneering Efforts of Detroit Police Commissioner George Edwards, 1962–1963, by Mary M. Stolberg, 1998

Toast of the Town: The Life and Times of Sunnie Wilson, by Sunnie Wilson with John Cohassey, 1998

These Men Have Seen Hard Service: The First Michigan Sharpshooters in the Civil War, by Raymond J. Herek, 1998

A Place for Summer: One Hundred Years at Michigan and Trumbull, by Richard Bak, 1998

Early Midwestern Travel Narratives: An Annotated Bibliography, 1634–1850, by Robert R. Hubach, 1998 (reprint)

All-American Anarchist: Joseph A. Labadie and the Labor Movement, by Carlotta R. Anderson, 1998

Michigan in the Novel, 1816–1996: An Annotated Bibliography, by Robert Beasecker, 1998

"Time by Moments Steals Away": The 1848 Journal of Ruth Douglass, by Robert L. Root, Jr., 1998

The Detroit Tigers: A Pictorial Celebration of the Greatest Players and Moments in Tigers' History, updated edition, by William M. Anderson, 1999

Father Abraham's Children: Michigan Episodes in the Civil War, by Frank B. Woodford, 1999 (reprint)

A Sailor's Logbook: A Season aboard Great Lakes Freighters, by Mark L. Thompson, 1999

Huron: The Seasons of a Great Lake, by Napier Shelton, 1999

Wonderful Power: The Story of Ancient Copper Working in the Lake Superior Basin, by Susan R. Martin, 1999